Dream Visions

Book Two of the Alien Dream Series

K.S. Riggin

TABLE OF CONTENTS

Prelude

In book one of the *Alien Dream Series*, **Bittersweet Dreams**, Frall, Clista, Clista's Uncle Cristoff, Franca, and Smog are being chased by an evil company named Cronoks Lab. That company is so desperate to get their hands on the powerful necklace that has physically adhered itself to Clista, they're even willing to kill.

They've already murdered both Clista's and Frall's parents in their greed to obtain all the Old Ones' artifacts, and their employees have traced Clista and her group to Smog's cabin, which lies out in the middle of a wilderness area.

Attempting to flee from these armed and dangerous criminals, Clista and her group hide in the woods, but Cronoks Lab somehow continues to track them, forcing the group to leave the trails and plunge deeper into the woods.

In the end, when Clista and her group finally reach the mountains where they've been drawn by their dreams and the urgings from the *Old Ones*, Clista and Frall meet up with a hologram named Thoric. He informs them that Clista has been chosen to become the next queen of the *Old Ones*.

Clista rejects such absurdity, but agrees to go to the *Old Ones*' planet, where she'll not only be safe from Cronoks Lab, but can meet the aliens and find out why they believe she should become their queen.

Thoric pressures Clista to marry Frall, but she tells him that she's too young and is not ready to commit. When Thoric realizes that he cannot force the mating, he agrees to take the entire group to Flagan Pren where the *Old Ones* currently reside. The hidden spaceship blasts off, and Thoric puts them all into deep sleep for a trip that will last approximately one Rwawan year.

In Book two, **Dream Visions**, the mystery of the *Old Ones* is partially revealed, but unfortunately, the *Old Ones* are nothing like the group expected, and neither is Flagan Pren.

Chapter One

I opened my eyes and realized that we'd landed on the planet Flagan Pren. Frall had just come in to kiss me good morning. It was a pleasant way to awaken, but I hoped he didn't expect me to look like Sleeping Beauty. My breath must be abominable, and I know my hair was probably a mad spider web of nasty.

Ms. Evans, who asked me to call her Franca, was in the bed next to mine. After a few minutes of conversation between Frall and me, we woke her up and went to greet the others. My uncle and Smog were in separate rooms, with Frall's room in between theirs.

Frall was dressed properly, but I was still wearing elegant blue pajamas, which I knew was not an appropriate outfit to go about in, even if I'd slipped a robe over them. But this was a new world, and I was too excited to have to wait for a first peek.

Just after we knocked on the men's door, Franca came out of the room. She was wearing a dress similar to the one we'd dined in for our last meal. She eyed my pajamas and sent me back to change. Frall laughed, but he didn't contradict her. I think he wanted her to remain on his good side.

Even though I shouldn't have, I skipped my morning shower and donned the silly blue dress that was laid out for me. It was a perfectly handsome dress, a nice shade of blue, and its fabric was soft and draped well. It's just that I was a jeans and t-shirt kind of girl. Perhaps

I'd be able to wear whatever I liked once everyone got acquainted with me.

I slipped into the sandals I found next to my dress and finger combed my hair, hoping to improve the bedhead thing. Then, I slipped out of my room for the second time and discovered that everyone was waiting for me. They'd gathered outside my door, yawning but ready to start the day.

After a quick good morning to everyone, which seemed oddly brief since we'd been asleep for so long, we moseyed down to the far end of the hall where a huge picture window would give us our first peek out at the new world.

We were all eager to gaze at an alien planet, the one that was to be our new home, but there was little to see: only a grasslike pasture. I guess it was grass, even though its color was tinged with blue. The hue was off just enough to make it look artificial and wrong.

We could see several trees, scattered here and there with branches bowed over like old men, and overhead, a sky the color of scummy lake water, except in the patches here and there that had sickly, orange highlights.

Our mouths dropped open, and I think we had the same thought. The view outside was of ugliness. What a disappointment after all our months and months of flight. I'd envisioned either pristine nature, like a tall forest, or a beautifully manicured city with garden plots of blooming plants and people walking about.

Thoric, the hologram who'd brought us to the planet, wasn't willing to discuss the gloomy appearance of the terrain or what would happen when we left the ship. In fact, he pretended that we hadn't even landed, wanting us to first awaken slowly, as he kept putting it. He insisted that we not meet anyone or leave the ship until we had fully recovered from our long passage through space.

So, with absolutely no discussion of our big disappointment at the view outside the spaceship, we followed Thoric to the great room where meals were served. It was there that it truly hit us that almost an entire year had passed since we'd last seen each other.

There was suddenly a rush of morning greetings. How did you sleep? Are you well? Did you dream, etc. I guess the shock to our bodies of such a long sleep was finally wearing off. So, although it felt as if we'd only spent one night in bed, there were long hugs mixed with the chattered answering of all those questions and a lot of expressed amazement that we were actually on an alien planet.

Poor Smog tried to step away from the outpourings. I guess he didn't really feel like he was part of it, but despite his protests, Franca and I hugged and kissed his scratchy face. When he tried to back away, we told him he was family and deserved to have to endure all the celebratory hoopla.

I centered next on my uncle, who'd flown across the stars for me, sacrificing his house, his job, and the life he'd left behind. I was very, very grateful to him, and I told him so. I'd thought of that many times during my long sleep. Had they been dreams of regret or feelings of guilt?

I knew that I owed Uncle Cristoff much more than I'd ever communicated. He'd come to take care of me after my parents were killed. He saved my life, too. I was sure the Cronoks Lab people would have caught me if he hadn't taken me under his wing. I needed to make up for being a moody brat back on Rwawan. I gave him a couple of fervent hugs, a heap of thank yous, and told him I loved him for the first time. He wiped his eyes and murmured the words back to me.

I thanked Franca as well. I was awfully indebted to her for her honesty and her guidance, and I couldn't believe that she'd been willing to leave everything behind to follow Cristoff into the unknown. That was amazingly romantic.

Last, of course, was Frall, but he and I had already had our own moments together. I'd shared private words of emotion with him while I was still in bed and he was sitting by my side, but I hugged him again for good measure and kissed his soft, warm lips.

Then Thoric offered us breakfast, so we sat down and ate and ate. We had pancakes, of course, since Frall and I loved them, and Smog had asked for steak and eggs. Uncle Cristoff wanted tomatoes and bacon. Franca asked for French toast. Breakfast ended up being a very large smorgasbord. Thoric also supplied us with cups and cups of hot coffee, somehow knowing that would be the most important part.

We rested that first day on Flagan Pren while we slowly processed the data injected into our minds during the long, deep sleep. We also practiced the language that had been placed inside our brains. The patterns, the verb forms, the vocabulary — that was all inside of us, but the sounds and the movement of our mouths — that was the part that was necessary to experience and rehearse. Our tongues curled slowly about the phrases and words. We stumbled frequently, stuttering and hearing the wrongness of our attempts, but by the end of the day it was already sounding better.

That night, when I lay down in the room that was becoming so familiar, I pondered what the next day would bring. My mind roamed through the many things I'd learned: the customs, the foods, the plants, the way to greet the *Old Ones* . . . I had a head full of dream visions, still fuzzy and maze-like. Some of them were as distant as if they belonged to someone else. Their lessons had no anchors yet, so they drifted about like leaves in a stream. I kept picking them up to inspect them, turning them to catch the hues and the patterns in their shape, but I couldn't read them. It was difficult to sleep that night.

Franca had fallen into her slumber almost the moment her head hit the pillow. I'd had no time to chat with her. I missed that. I closed my eyes with a long drawn out sigh and wondered about meeting the *Old Ones.* What would they be like? Would they answer my questions? Would they explain why they thought I should be their queen?

I don't know how long it took for sleep to come, but I do know I that when I was finally swept up into it, I dreamed of Frall.

The next morning, we woke up eagerly, ready to meet the *Old Ones* and to finally leave the ship. I couldn't wait to see more of the planet. There had to be something beyond sick-looking grass and stunted trees. Flagan Pren was supposed to be as big as Rwawan. Maybe we'd landed far away from all the cities. Maybe this was just the landing spot for spaceships, although I hadn't seen any others.

Unfortunately, Thoric was still not convinced that we were sufficiently rested, so we had more time to wander endlessly about the ship and eat with unlimited choices in its dining room. We stayed together in our explorations, finding a room full of marvelous picture books, all of which we could read. Apparently the sleep tapes had covered the written language as well as the spoken. There were other kinds of books, as well, but remaining in a library when we were restless to explore, didn't merit much lengthy reading.

It wasn't until late afternoon of that second day that we were allowed to proceed with our introductions. We clothed ourselves in fresh garments, those which reflected the customs of the *Old Ones*, which meant long gowns for Franca and me and tunics and trousers for the men. Our shoes, though, were a startling component, for they had four inch soles and made us feel like we bounced as we walked. Franca's and mine were coordinated with our dress colors, but the men wore a dark hue, a deep, rich bronze. All of the boot-like shoes were tied with closures that fastened with a click and adjusted to our feet as we moved. I must admi that the shoes were very comfortable.

As we approached the ship's lower level, outward port, it yawned open, and we stepped down onto the blue-green lawn. Thoric told us the grass-like plants were filled with tiny organisms that exhaled oxygen into the air. I picked up a handful of them to see their texture. They felt soft as tree lichen. I lifted them up for a good sniff, but Thoric, suddenly noticing what I was doing, knocked them out of my hand.

"Please, do not touch the plants until your body has had time to adjust to Flagan Pren. There is nothing that *should* harm you, my queen, but it is wiser, always, to explore slowly."

"I agree, Clista. Be cautious," my uncle said. "You have no immunities here. It's like we're newborn babies, except babies carry their mothers' resistance to disease. We don't."

I nodded. "I'm sorry," I said as Thoric wiped my hands with a damp cloth. Smog shook his head at me and made a gesture about my not using my brain.

"Leave her be," Frall snapped. Smog rolled his eyes and glanced over at my uncle, exchanging a look that must have held some meaning I didn't understand.

We were heading towards a high tower, a very beautiful one that reflected light from the second, slightly reddish sun. Small points of brightness prismed into rainbows and shot out at different angles, making the tower seem more like a decorative ornament than the building Thoric claimed it to be. It formed the center of our attention as we strolled, for not only was it magnificent, but there was absolutely nothing else to see.

The path we trod on was cushioned deeply by the growing matter I'd tried to smell. Each stride of ours stirred up an odor similar to dill weed. I wondered if we should be wearing facemasks if the plant were really so dangerous, but I didn't say anything. Thoric probably would have whipped some out.

The land around us stretched out so flat it seemed as if someone had ironed out the wrinkles. The sea green of the moss-like plants extended far into the horizon and showed only a scattering of trees. The few trees that did manage to argue with the grass were very different from any I was familiar with. They had leaves and branches that hung in thick, frilly screens that completely draped their trunks, as if seeking to hide their trees' lower bodies.

Except for the quiet, scuffing noises of our bouncy shoes, everything was silent. No birds called. No tree leaves rustled. The hushed manner of the area reminded me of the forest's quiet. On our trek towards the mountain where Thoric had waited for us, each day before the sun rose in the sky had held that same moist stillness.

Where are all the *Old Ones*?" I asked.

"We are not as numerous as you Rwawans, my queen. In Flagan Pren, there are only 192 citizens, and most of them flutter from planet to planet. The *Old Ones*, as you call us, prefer the quiet. They rarely assemble beyond two or three people."

Nothing on the alien planet was like I'd thought it would be. No cities full of lights. No spaceships zooming off into the sky, heading for other galaxies. No strange creatures nodding antennae and nine-fingered hands. I sighed again, kicked at the dirt and the blue-green plant life, and glanced at Frall, who wasn't looking my way because his eyes, like everyone else's, were centered on our destination, the one building that stood amid this endless sea of nothing.

I'll admit that the structure became even more interesting as we neared. Its cylindrical shape rose higher than even one of Rwawan's giant sonar beacons. The building's dark shadowed windows formed numerous colorful lights that reflected from the planet's two suns.

"It's really beautiful," I said again, although I'd probably already said that a half dozen times.

"Why would the *Old Ones* make a building this large if no one wants to assemble in groups?" Smog asked, as if he'd been thinking over the matter for a while. "And what's the point of making it multi-leveled when you have all this space around?"

"Ah, merely for the splendor," Thoric answered, smiling proudly as he stared up at it.

"It's definitely a work of art," Franca said, stopping to admire it.

"Doesn't make sense when there's no one out here to appreciate it," Smog grumbled.

"You once said that you didn't expect company at your cabin," I reminded him. "But, yet you decorated it in natural woods and . . ."

"That's not the same thing at all," Smog muttered. "That was just to please myself. Besides, it was convenient to use materials at hand."

Frall glanced over at him and snickered. "You put in inlaid solid wood floors for convenience? Cement and plastic are a lot faster and easier to work with than wood."

"Knock it off, Frall," my uncle warned him, still playing the heavy.

I quickly changed the topic. "Thoric, when do we get to meet you in person?"

The hologram stifled a chuckle and met my eyes. He'd already commented several times about my continued role as the peacemaker of the group. His eyes reminded me of the fact.

"You will see me as soon as you enter the building, my queen. This is my home," he told us.

Thoric was truly an amazing hologram. He registered pride just as visibly as any human. His face practically glowed with it.

"I'm glad," I said. "Will I be able to hug you then as a thank you for your kindness?"

"Kindness? Isn't he the one that slapped that necklace around your neck and saddled you with all this?" my uncle growled.

Thoric turned to him and bowed slightly. "I did not make the necklace, Cristoff. My ancestors did. The Vremorel, as it is referred to officially, was buried in what you called the Stormian Caves for over a thousand years. I am old, but not that old."

"Could we leave this conversation for later?" Smog asked. "If we're going to get there, we need to keep moving forward."

No one commented about Smog's point, but we all resumed our longish hike, our eyes engrossed in the structure that towered over us so majestically.

"How old are you?" Frall asked Thoric, as my boyfriend swung an arm around my shoulders.

I didn't mind when Frall did that. In fact, I rather liked it, except my uncle had been glaring at us ever since he'd learned that Frall and I had awakened earlier than he did. I wasn't sure what my uncle thought had happened, but I'd gotten a lecture about it already. I was supposed to be keeping Frall at a distance so he didn't get any "more ideas."

"Uncle Cristoff won't like this, Frall. He ordered me to keep away from you," I whispered to Frall.

Unfortunately, even that brief chat was misconstrued.

"Come here, Clista," Uncle Cristoff barked. "I want you to walk beside me for a while, please."

I sighed. My uncle was being a real pain. His constant nagging challenged all my resolutions not to flare up at him again. But I remembered how he'd been spurned by my mother, the apparent love of his life. I guessed that he had the right to be a grouchy, embittered man. Sometimes.

Despite Frall's look, I pulled away from him and waited for Franca and my uncle to catch up with me. My smile was probably a little forced, but I did my best. Sometimes, adults can be very tedious.

"Thank you, my dear," my uncle said. "I thought we had an agreement . . ."

"Oh, look!" Franca called out, interrupting what was starting off as another lecture. "The archway is changing!"

I thought that Franca was just smoothing over the tension, but when I looked up, I saw what she was talking about. "Whoa! That's incredible!"

The archway was entirely remolding itself. There's no other way to put it. It was like metal melting and sliding into a liquid state, then reformatting to become an open door, one with a sparkling black pathway flowing out to welcome us. Overhead, the archway was enlarging, too, forming curlicues of charpas and dafnines, the native wild flowers of Rwawan. As we watched, color poured into the spiral figures, and they opened into huge, vibrant blooms.

"Stars and comets," Smog said, "Now all it needs is the scent."

No sooner had Smog said that, than a gentle breeze stirred, and the fragrance of fresh flowers rode the surrounding air.

"Oh," said Franca. "How heavenly! I adore the scent of charpas."

My uncle frowned. "What scent? I don't smell a thing."

The rest of us turned to stare at him. "Their perfume is so strong we can almost taste it," I said, not believing Uncle Cristoff couldn't.

I breathed in again and then again, admiring the flowery fragrance.

Frall sneered at my uncle. "It figures that he can't smell it."

My uncle gave Frall a look that said he'd better shut up. The looks passing between them could have singed paper.

I sighed. Would the two of them never make peace?

Frall glanced at me for some reason and muttered, "Sorry."

"Maybe you're coming down with a cold," I suggested to my uncle. Frall laughed, but he tried to cover it up. I hoped he'd quit jabbing at my uncle. It would sure make things easier.

We were at the entrance of the gigantic tower by then. Smog took a step onto the hardened floor, which looked exactly like obsidian: shiny, black, and smooth. "Seems okay to walk on, folks," he said as he continued on.

Frall was next. He stepped up onto the slight, stair-like rise and turned to face us. "This is cool, Thoric. What other surprises are you going to throw at us?"

My uncle, Franca, and I brought up the rear. I stepped onto the surface rather hesitantly, wondering what would happen if the rock started reformatting while we were on it. Thoric glanced back at me and smiled, then shook his head.

As hard as the floor's surface felt, it was strange that our shoes made no sound as we walked. I stared down at its glasslike surface. Frall stopped to watch us for a moment, his eyes on me, before brushing over my uncle. Then he turned abruptly and hurried off to catch up with Smog, who was still striding forward with all the confidence in the world.

"There will be many such treats for you, Frall," Thoric called out to him as we walked on. "This particular technology is customary here, but you may find our technology and architectural styles quite different from what you're used to, perhaps, even frightening at times. But please, remember, nothing here will cause you harm."

Frall didn't comment. He seemed more intent on keeping up with Smog. It was as if it was important to him to be the first to enter the building. I thought, actually, that Thoric should have been leading, but he always seemed to walk near me, not at my side exactly, but wherever he could keep an eye on me. I rather resented that. I had no intention of picking up any more strange plants, and hadn't he just said nothing would injure us?

Frall and Smog passed under the flowered archway and on into a small vestibule — at least, that's what they said it was when they called back to us. We three, slightly dragging and at the back of the line, stepped inside and halted, amazed at the walls all around us. For one thing, they were plain — no pictures, no decorations, nothing, and, for another, the vestibule didn't seem to go anywhere.

Almost the moment we entered the cylinder-shaped room, the outer door melted inward and then closed off. It felt like we'd been imprisoned in an old-fashioned soup can.

"Oh, no," cried Franca. "I don't like this."

The rest of us didn't much care for it either. We turned to look at Thoric, to ask him about it. Unfortunately, he'd chosen that moment to blink out.

Chapter Two

My uncle wrapped one arm around me and around Franca and pushed us together. "You two stay close in the center here. I'm going to knock on some walls.

"Frall, Smog, come on. Let's see if we can make a hole in this."

The men approached the wall and slammed their fists against it, having nothing harder than that to use. Their hands merely bounced off. They repeated this several time. Frall even used a couple of karate kicks, but nothing even dented the wall. Smog stopped to shake his hand, like his last punch had brought pain.

"Are you okay?" I cried out.

"You want to kiss it for me?"

"Hey! My hand hurts, too," Frall whined.

"Look!" Franca yelled, "The wall is dissolving!"

I turned around, as did the others. The far side of the wall was now flowing like hot metal, blending into a river of slowly moving reformation.

"It's making a door," Frall said. "I knew the *Old Ones* wouldn't let us down. It's just making a door."

"If you knew that was true, then why didn't you say it before I broke my hand on that stupid wall?" Smog complained.

"Did you really?" I asked.

My uncle laughed. "He's too tough to break anything, Clista. Don't worry about him." Uncle Cristoff walked over to us and swung his arm around Franca. "You okay?" he asked, as if she and I had done anything other than stand there and watch.

Franca smiled up at him. For some reason, it made me nervous to watch them. I looked away and discovered Frall staring at me. He gestured for me to come stand with him, but I shook my head and nodded at my uncle. Frall shrugged, giving me such an adorable smile I almost melted.

"I love you," he mouthed.

I was just about to answer him when the door twisted itself open. "Come on," my uncle said, taking the lead.

We walked forward into a huge room, one almost exactly like the throne room onboard the ship. The same massive stone chairs sat atop a raised platform, a repeat of the decorations adorned the wall, and the marble flooring with its gray rivers of swirl spread out just like before, shining and elegant. But the room was deserted. We exchanged glances, wondering if we'd somehow returned to the chamber back in the spaceship, where we'd first met Thoric.

Then a voice boomed out. "Sit down, Queen Clista and Frall. The thrones belong to you. Chairs will be brought for the others."

"I'll stand," I said, moving closer to Frall. I suddenly couldn't worry about pleasing my uncle anymore. I needed to be next to Frall.

"Clista," my uncle said, demanding my return, but Frall, taking his cue from either my proximity or from something the *Old Ones* were whispering into his ear, surrounded me with his arms. Then he touched my neck with his lips, and I felt the warm breath of his words whispering the magic of the *Old Ones*.

"What are you doing?" I cried out, attempting to break free from his spell.

"Only telling you secrets, Clista. Listen."

The rest of my group stood still, not moving, their eyes directed forward. I suspected the *Old Ones* had placed them in stasis again, as they'd done the night when Frall and I had first entered the cave.

"Frall. Please don't…"

"You know I'd never hurt you, Clista. I can see your fear. It's in your mind. You think I'd use any magic that would hurt you? I'd never do that."

"Then let me go, Frall."

He sighed and opened his arms. "I was only supposed to talk you into sitting on the throne, Clista. I didn't mean to frighten you."

I turned around to look him in the eye. "Then why did the *Old Ones* freeze my uncle and the others? Look at them, Frall. Don't you see that they're up to something?"

"No, not at all, my dear," said a voice very much like Thoric's but deeper and richer. I turned to look.

"Your mind is overly suspicious, child. Sit down, Frall. I will talk to her myself."

This was the true Thoric, not our friendly little hologram. I was almost positive, but still, I asked. "You're real, aren't you?"

"Yes, Clista. I am real. I am Thoric of the ones you call the *Old Ones*."

"Why have you done all this? It's you who've made me distrustful. Why did you trap us in that small room and take Thoric away so he couldn't explain what …"

"Questions, questions, questions. You always have questions. Be patient, my queen. We want to help you and to do what is best for you, but you have to trust us."

The *Old One* looked exactly like our hologram Thoric, but yet, there was an air of substance or vitality in him that the hologram hadn't possessed. I studied this new Thoric, feeling the déjà vu of knowing him, yet not having a lot of confidence in this new version.

"Trust?" I said. "You've frozen my uncle and the others. You've twisted Frall . . ."

"Twisted?" Frall said, lurching up from his seat on the throne. His face looked crushed, as if I'd said something horrible about him, but I hadn't meant it that way. Not exactly, anyway.

"Who are you calling twisted?" Frall repeated.

Thoric gestured to Frall to sit down. Frall dropped back down, but he continued to glare at me. I could see out of the corner of my eye that he was angry, but I didn't have time to deal with him. It was Thoric who demanded center stage.

"Do not be concerned with your family, Clista. I will free them shortly, but it is better if you and I meet and talk privately for a moment."

I didn't say anything. I remembered how I'd thought I'd want to hug the real Thoric and thank him when I met him, but that didn't seem appropriate anymore. How could I trust someone who played with us, terrifying us with power games?

"Clista, my queen, I am not playing with you, not at all. What I did was necessary. I enclosed you in that small room for decontamination reasons. That is a safety feature of every house for people who fly from planet to planet. It did not occur to me that the process would frighten you. I have told you that no one will hurt you. Did you not believe me?"

"Stay out of my mind," I said.

"Clista!" Frall said. "Have you gone mad? Thoric is one of the *Old Ones*. Don't speak so . . ."

Once more, Thoric waved Frall silent. "The queen of the Brith has always been this way, Frall. She must be allowed to question. We must allow her to speak freely."

I hated that Frall and Thoric seemed so familiar with each other. I felt left out, like I'd missed half the conversation. Had Thoric put me into stasis for a while so he and Frall could chat? I turned my back to Frall and focused on the *Old One*.

Thoric smiled like we were friends. Obviously, he was feeling quite comfortable about the situation, but I suffered from being lost in it. It wasn't like not having done my homework correctly, but I felt like I was back in math class with bilateral equations swimming upside down and Bista nowhere close by to help.

"Patience, Clista. All will be made known to you, and the "stasis," as you call it, is inoperative on you and Frall. Your Brith blood makes you immune."

"Brith?" I repeated stupidly.

"Yes, but I was explaining about the entrance chamber. During the time when you and the others were sealed inside, the holographic light, Thoric as you have come to know him, was invalidated. Holograms do not function during the contamination process. The absence of the one you trusted frightened you, and I am sorry for that. I did tell you repeatedly that no harm would come to you."

I closed my eyes, trying to make sense of it all. I let go of the odds and ends that Thoric kept throwing at me. I couldn't process everything at once. I just concentrated on thinking about one item — the decontamination chamber. I suppose the way Thoric told it, it did sound logical, but what about Frall? Why did he seem so relaxed and sure of Thoric? Why wasn't he suspicious about the chamber and about . . .

"Frall and I have been communicating throughout the days of your journey, Clista. He and I are friends already. He understands the whole. He knew I would never harm you."

I knew the *Old Ones* had sent dreams full of information to Frall while we were in the forests of Rwawan. Frall had always seemed to have information I didn't.

"Why do you keep pushing Frall and me together?" The question burst out of me as if it was the biggest ambiguity of all. I hadn't meant to ask it, not before I'd cleared up all the other mysteries, but that question rose up of its own accord, taking prominence over all the others.

Thoric laughed. His chuckles rippled through the hall and echoed back to me. Then he sobered and grinned wider. "Ah, Clista. It was my fault that Frall took you into his arms in the hall. That upset you, and I'm sorry. I asked him to do so because I thought it would relieve your fears to feel his arms around you. It was meant to make you feel better, not to alienate you from his touch. He, of course, was eager to hold you. I read no reservations in his mind.

"Forgive me for my miscalculation, but you do need such contact, Clista. We all function better with our mates nearby. The Brith, you see, are always paired.

"And you love him, Clista, although, for some reason, you fight that. Look into Frall's eyes and tell me that you do not feel his love. He yearns to hold you. Tell me that you do not feel the same."

"Do you have to rub it in?" Frall asked, but his eyes on me confirmed Thoric's statement.

None of that was new. I'd been aware for a long time how Frall felt. I just wasn't sure if I returned those feelings, or at least if I returned them as strongly.

Thoric chuckled. "Yes, you fight the connection. That is your mother's training. She also would not submit to love. She taught you to analyze instead of to feel."

"Don't speak of my mother that way. She…"

"I have not maligned her, child. She brings to you your Brith blood."

I'd listened patiently to every single thing Thoric said. I really wanted to understand it, but he was confusing me again.

"Brith blood? My mother?" I rasped out, as if the words were choking me.

"Stop a moment, please. I don't want to talk about this. Please. Just give me time. Leave me alone until I can make sense out of everything."

I shot a glance back at Frall. His eyes were fastened on me. They were no longer angry. In fact, he had a feather of a smile on his lips, the same look he'd worn the day in the gym when he'd exercised beside me and meddled with my report.

That memory took me back in time, seeing myself frowning because he'd interrupted my research. I wished we were back there now, still at that moment, a moment when my parents were alive, and my biggest problem was working with an irksome but incredibly handsome Frall.

But one can't go back in time. I knew that. I breathed in and out slowly and pulled the threads of my memory back inside me, reminding myself that I was on Flagan Pren now and needed to stay calm.

"Okay," I said, swallowing hard, but determined to get to the bottom of all this. It was hard not to glare at the alien, though. I felt that same itchy, scratchy rage I'd once felt with Frall.

I drew in a full breath and held it a minute. "I understand, maybe, about the entrance room, but not about the Brith. What does that have to do with Frall and me? Why do you say my mother…"

"So many questions, my queen. I do have the answers for you, all the answers you require, and so does Frall. Pity you never listen to him. He tries to tell you, but his responses make you fearful. Why would it be any different if the information came from me? I think you are not ready."

I sighed. One of the things I hated more than anything else was evasion. Thoric was dodging. "Please. We've traveled so far, and you're talking in riddles. I don't understand. I don't shut Frall out. It isn't like that at all. But just tell me first. What do you want with me?"

"Another question, and yet I have not answered your first one. Sit down, Clista. If you sit, I will reply to anything you ask — anything in my power to tell you."

Instead of giving in to the request, I backed away. "Why are you always trying to get me to sit on that throne? I won't, you know. All this queen business is silly."

Thoric sighed. He raised up his hands. His long tunic rearranged itself, just like the metal had, flowing and rippling to fold itself neatly.

He clapped his hands once, and a hologram appeared. It wasn't the one that looked like Thoric.

"Chairs for all," Thoric said.

Then he shook his head and smiled at me. "Sit on Frall's lap, sit on your throne, sit on the chair that is brought to you, even stand — it doesn't matter, Clista. I have no nefarious schemes. There is no evil in this room. I simply greet you as a fellow member of my race."

"What are you talking about now? A member of *your* race? You mean, like an honorary one?"

I glanced at Frall, wondering if he really did have all the answers. He wasn't looking at me, though. He was sitting back, his feet stretched out, his eyes closed as if he were listening to music or words that someone was whispering into his ears.

"Frall," I said. "Are you okay?"

His eyes shot down to mine. He laughed. "Yes, Clista. I'm fine. I'm home."

"Get out of the throne," I screamed. "They're doing something strange to you. I don't know what — brainwashing or magic or …"

Thoric laughed. "Frall just knows the answers, Clista. He's waiting for you to know them, too."

A hologram I'd never seen before brought in chairs. He placed them beside my family and friends.

"Shall I ask him to bring you a chair?" Thoric asked.

"What about you? Why don't you sit on the throne, and I'll take the chair he has brought for you," I suggested, not trusting Thoric's tricks.

He laughed. In fact, he laughed so hard tears sprang into his eyes. Then he dropped down into one of the newly brought chairs. "Bring another one," he ordered the man. "Queen Clista, you are truly Brith."

Chapter Three

I was silent, watching him, distrustful of anything he told us and of the way he'd given in so easily about the chairs. Yet, when the hologram returned, bringing me one of the chairs, I sat.

"First of all, that man is not a hologram, Clista. Torral is of another race, a race which serves me extremely well, for they have no speech. Constant questions would eventually become tiresome, wouldn't they, child."

Another race? The man is from *another* race? But he looks like us — like the *Old One*. How was that possible? My brain felt overloaded. I was going down for the third time. Gurgle, gurgle, gurgle. Fried circuits. Smoke. Fire. Really not coping well. Thank the stars my Academy proctor was not here to witness it.

"On the ship, you sat in my lap, Clista," Frall said. "Please. Let me support you. Please trust me. Remember what we've been through together. Please."

I felt like crying. The quicksand I was standing in was swallowing my legs. "I *can't* trust you anymore, Frall. You're doing what they tell you to do. It's not *you* anymore. I'm sorry." I told him brutally but honestly.

Frall reacted predictably. He bolted up, stepped down from the platform, and started to walk away. Then he stopped and turned to

Thoric. "No. You don't understand. There's only so much I can put up with. I have to get out of here before I explode. Let me go."

Thoric looked thoughtful a moment. I think he was saying something else to Frall via mind to mind. At first Frall shook his head as if refusing, but after a moment, he sighed and nodded.

"I am glad you're willing to stay for a while longer. Clista doesn't know it, but she needs you." Thoric said, not bothering this time to hide his words from me, although it was pretty easy to see that he'd been mind-talking before.

Frall glanced at me, his eyes shadowed. My heart pounded. I wanted to run to him. I wanted to comfort him, to allow him to take me into his arms, but I didn't. I gripped my chair and held myself still.

"Stubbornness is tenacity. Remember that, Frall. The very qualities she is demonstrating now will serve the kingdoms well. Observe her slow analysis, her caution in making a commitment, and her integrity. Be patient with her. She will be your mate soon enough. You know that."

"STOP SAYING THAT," I cried out, springing from my chair. "You have no right. Why am I here? Answer me that one question, then, no more for today. Just tell me, why am I here?"

"You have wasted your question, child, for you already know the answer to that one. We must train you to rule. You must learn of your history and that of your people. You must…"

"You keep saying that. Okay, I wear the necklace. But only because you put it on me. I don't understand why. There's nothing special about me. I'm just an Academy student, a normal person, or, at least, I used to be until all of this stuff started happening. Why are you trying to confuse me?"

"You said no more questions. So our discussion must be over."

Thoric clapped once, and Smog, Franca, and my uncle blinked their eyes. Robotically, they each sat down in the chair beside them.

"What happened?" Smog asked. "What . . .?"

"Welcome Rwawans. Welcome to Flagan Pren," Thoric said, his eyes scanning them, looking into each one's eyes.

I ran to my uncle, dropped down to the floor on my knees, and threw my arms around him. I knew I was retreating into childhood, but I needed someone older, someone reliable. Maybe, at times, Uncle Cristoff was a pain, but at least I trusted him.

"What's wrong?" he asked me. "Frall again?"

I shook my head. I needed to explain everything to him but where to begin?

"What did you do to her?" Smog yelled, launching himself out of his chair like he was about to attack Frall.

The voice we'd heard at chamber's entrance boomed out through the walls, "Be silent and sit."

I didn't move. I stayed, clinging to my uncle, but everyone else froze in place except for their eyes. All eyes moved to Thoric, waiting to see what he'd do next.

"Clista, you prefer to be at the feet of that man? I have offered you the throne of your heritage. I even brought you a simpler chair when you requested it. I have given you a mate whose reassurance you are currently spurning. What is it you want, my queen?"

I wanted to scream or to run and hide. Either of them or both would have been preferable to the confusion I felt when I listened to Thoric's constant flow of riddles and puzzles.

Uncle Cristoff stood up. Gently, he lifted me and placed me down in his chair. "Sit with Franca, Clista. Let me talk with this guy."

Franca scooted her chair closer, then put her arm around me. I welcomed the companionship, but my eyes were on my uncle.

"I think I'll mosey on over there, too, ladies," Smog said, but he didn't. He was suddenly stuck in his chair. He struggled against the semi-stasis, calling out to my uncle. "Hey, buddy, looks like the old guy is up to some pretty mean tricks. I can't move. Guess you're on your own."

"Thoric, let him go," I yelled.

"Is that a command, my queen?" Thoric asked, his amusement causing crinkle lines under his eyes.

I sputtered, looked back at Smog, and turned my head to glare at Thoric. "Yes. Let Smog go."

I looked back to see if he was loose. I couldn't tell. He'd disappeared. My uncle was once more frozen. Franca wasn't moving either. I blasted out of my chair. "What are you doing? Where's Smog?" I cried out. "Frall, what's going on?"

Frall glanced over at Thoric. "You're pushing her too hard. You should back off."

Thoric considered the suggestion. "All right," he said. "I believe you know her better than I do. That will be enough for today. Wander about as you please. You will find rooms down the hall on your left. Food is to your right. I will see you all tomorrow. Rest well."

"But, what about. . ."

Thoric looked back at me. Then he let a great roar of a laugh. I blinked, and he faded out.

"Where did he go?" my uncle questioned.

Smog yelled out that he'd gotten free of the chair. Franca questioned me about something, but I was staring at Frall. "Will you talk to me, please?" I asked.

He smiled, but it was a faraway look. Then his eyes met mine, sending shivers down my spine. His eyes looked so knowing, like Thoric's, in a way. "You don't trust me. Why should I?"

"I want to trust you, Frall. I really do. I want to understand everything."

He shook his head. His face now just looked closed off, his eyes shuttered from me. It was like that time when I told him about being captive to the necklace. He was rejecting me, rejecting my words.

"No. I don't think so," he said. "Remember in class, Clista? We had to work together. We were learning how to do that, but we never finished those sessions. That was really hard for you, wasn't it? This time, I'm on the same team. I'm not obstructing you. No proctor has given me an assignment to cause trouble. I'm on your side, Clista, but you're still treating me like I'm your enemy.

"I won't go to you again, Clista. Thoric keeps urging me to do so, but it doesn't do any good. You just run away faster. I've bent over backward too many times. No more. You'll have to come to me," he said. Then he turned and walked away.

Chapter Four

"What was that all about?" Smog said, whistling through his teeth. "How'd you get him in such a huff?"

"Frall," I called out, ignoring Smog, but Frall didn't stop. He turned the corner and disappeared from sight.

Franca, Smog, and my uncle began talking all at once. They were smothering me with words. Their eyes drilled into me, and their hands reached out. I could hear the buzzing of their voices, but I couldn't understand them. I shut my eyes to close out the sight. The vision of Frall's back and his rebuff played itself over and over in my mind. I dropped down into my chair and covered my ears. That's all I remember, tuning it all out and going inside.

At some point, I heard Franca say, "She's in shock. She needs quiet." The meaning of her words trickled in, but I ignored it. I was still focused on the images in my mind — collating, assessing, and trying to process them.

Someone placed me on the ground. They put my uncle's jacket over me. The smell of him was on it. I breathed it in. It was a nice smell, but it was different than Frall's. A ripple of horror hit me. Once again, I saw Frall walking away.

Franca had placed her sweater under my head. It was soft and a little itchy against my neck. I remembered Frall's warm breath as he lay beside me. So many days he had been at my back — his quiet breathing, even sometimes, his snores. I missed that. I missed Frall.

I remembered things we'd done together, the way we'd traveled hand in hand, the kisses, the laughter, and the conversations we'd exchanged. I didn't open my eyes as I lay on the ground, remembering all those times, thinking back.

At some point, Smog came over to stare down at me. I opened my eyes to look at him. He bent down and said, "Open."

He slipped a grape into my mouth. I smiled and sat up. "You found the food," I said, swallowing.

He sat on the ground beside me. "Your uncle and Franca went to eat. They've been watching over you for about an hour. Are you better?"

I nodded. I didn't want to talk about it. I'd probably worried them, and I was sorry, but I was still too confused about everything. I still hadn't sorted it out yet. I should have if I'd really been lying there an hour, as Smog said, but all I'd been able to think about was Frall.

"Feel like getting up and getting something to eat?" Smog asked, poking another grape into my mouth.

"Stop it," I said, turning my head away. "I don't need to be fed."

"Good. That sounds like the Clista, I know." He popped the grape into his own mouth and chewed, still watching me as if he thought I was about to do something — like turn into a babbling idiot or erupt into violence.

Then, apparently discarding those ideas, he yawned, stood up, and reached a hand down. I took it and pulled myself up. I started to pick up my uncle's jacket and Franca's sweater, but Smog stopped me. "Ah, leave them, Clista. Who's going to steal anything?"

I guess he had a point. I left the clothing on the ground.

"Have you seen Frall?" I asked.

"Nope. But that's okay. That kid's been getting under my skin lately. He's too close to the *Old Ones* for my liking."

"I know. I think so, too, but…" I stopped. I had no way to conclude the sentence. I didn't know what came next.

Franca and my uncle were sitting at a very large, round table covered in sky-blue velvet and set with platters of fruit and sandwiches. A pitcher of what looked like lemonade was the centerpiece of a set of crystal glasses. The table and the comfortable-looking chairs placed around it were the only furniture in the large white-walled room.

"Wow. How pretty," I was referring to the elegant tablecloth, but the food attracted my eyes as well. There were purple grapes like Smog had been eating. Apples, oranges, pears, and a small melon-looking fruit I was unfamiliar with took up the rest of a silver platter in the middle of the table. Sandwiches piled high in staggered piles of dark bread and wheat. It looked like they had fillings with an assortment of ingredients. My stomach growled.

Franca rose from her chair, came over to me, and placed her hand on my forehead. How motherly she was becoming. I put up with it, even though I knew I wasn't sick. I was just on overload — brain sunburn.

When she was content that I wasn't feverish, she kissed my cheek. I thanked her for her sweater pillow and apologized for zapping out on them.

"Zapping?" my uncle questioned. Then he laughed. "Sit down before you zap out again. The food's pretty good. We've probably got whatever you want already on the table, but name anything, and it appears a few minutes later."

"How do they do that?" I asked, sitting down and reaching for a bright red apple.

"Wish I knew," Smog muttered. "I'd install it in my . . ." He stopped then, just realizing, I guess, like the rest of us, that his cabin might not even have a kitchen anymore. What had Cronoks Lab done to it? What had they done to all of us?

The apple was delicious, juicy, and sweet. I hadn't eaten one in a long time. The last time I'd had one had been when Frall brought it to me. I sighed and placed the apple down on my plate.

"Did Frall come and eat something?" I asked, looking about as if he were hiding in the corners of the room.

"No, we haven't seen him," Franca said gently, throwing a look at the others that I didn't understand.

"Eat a sandwich," said my uncle. "That boy will come back when he's hungry."

"But where did he go? Do you think he's just wandering around inside this building, or would he go outside?"

"Shoot, he could be with Thoric for all we know. That's probably where he is, in fact," Smog said.

"I *was* with him a few minutes ago."

We all turned. When had Frall entered the room? It was obvious he'd overheard our conversation.

"Are you all right?" I asked.

"What does that mean?" he questioned bitterly. "Did I survive your rejection once again, you mean?"

"That's enough of that, young man," my uncle snapped. I swung my head to look at him. Yep, his eye had a twitch again, same as my father's had when something irritated him.

I ignored him anyway. "Frall!" I said, standing up and walking toward him.

"Sit down, Clista," my uncle ordered. "Don't feel guilty for doing what's right."

I stopped. Frall's eyes were drilling holes into me. I wanted to please my uncle. But I wanted to please Frall, too. Why couldn't I do both? Why couldn't Frall be content with just being . . . my what — a friend? A boyfriend? Something like that.

But he was more than that. Our connection ran deeper than that of a mere school friend or even a steady boyfriend. Deeper feelings were stirring inside me. I loved him, but I also knew that I couldn't handle anything more. Not at the moment.

"Please, Frall, please, come and eat something," I said, my voice cracking on the last word as I gritted my teeth to keep from breaking into tears.

My uncle was saying something. Franca, I think, was speaking, too, but I shut them out. I was listening for Frall's answer. He was the only one who mattered at that moment.

He took a step toward me, his eyes still staring into mine. *Darn it, Clista. Why do you have to be so beautiful? And so innocently sweet?*

He didn't say that out loud for the others. He spoke it directly into my mind. I gasped.

"What's wrong?" my uncle demanded so sharply it broke through my concentration.

"Oh, leave them alone, Cristoff. Let them be," Franca said, pulling my uncle up out of his seat. She tugged him out of the room, and after a moment of silence with Frall standing frozenly stiff and me just staring into his eyes, Smog trailed after them. I hardly noticed. I was entrapped by the look on Frall's face and the whispers he was speaking into my mind.

I want to be angry with you, Clista. I want to leave behind the hurt and the strength of these feelings I have for you, but I can't. You're always there. You're in my mind, even when you're not in front of me.

You were thinking about me earlier, weren't you? Do you know what that's like? Every memory you think about hits me. I saw your tears in my mind. I felt how my words were like arrows, causing you pain. I'm sorry, Clista. I really don't want to hurt you.

You were thinking about us when we were in the forest, weren't you? How we walked hand in hand, the stories we told each other, the way I used to find violets hidden under the ferns, just to show you . . .

I'd told myself I wouldn't cry, but I couldn't help it. I was suddenly running toward him, flying into his arms. "I'm so sorry," I said. "I don't know why I didn't trust you. I do trust you, Frall. I do."

I know, Clista. I know.

Our lips ended our apologies. We kissed as if we hadn't seen each other for months or years. It seemed that long. Being at odds with the one you love is the most miserable thing I can think of.

Then, finally, after my tears were shed and our kisses grew less anxious and began to change in nature, we stopped and walked to the table. Then we ate, laughing and talking like good friends again.

It was so wonderful to have Frall back. I couldn't help grinning like a cocoa bear. I kept springing up to hug him, hardly letting him take a bite.

"Clista, you need to eat something, too," he reminded me the third time I did it.

I sighed. I guess he was right that I was overdoing the hugs a bit, but I'd missed him so much. He scooted over in his chair and chuckled as he kissed my cheek. "Eat a sandwich, and I'll kiss you again," he promised.

After lunch, Frall showed me the things he'd scouted out. He led me to an enormous inner orchard on the top floor. It was filled with trees loaded with fruits of various stages of ripeness. Several robots were harvesting apples and pears. They ignored us, except when we spoke to them. Then they stopped and turned to face us. The little lights in their eye sockets changed from blue to green each time they answered. I was surprised that they spoke our language, but Frall told me that he'd asked about it. They spoke all the languages of the Flagan Pren.

"Where is Thoric?" I asked one of the robots.

Its green lights blinked on and off three times, and then it said, "Thoric is on the third floor."

"What's he doing?" I wanted to know.

"Thoric is talking to his life partner."

"His life partner?" I hesitated. "You mean he's married?"

"He told you that, Clista," Frall interrupted.

"Yes, Thoric is married," said the robotic voice as if to reiterate what Frall had just said.

"Wow! I didn't realize that. I thought he lived here all alone."

Frall looked at me funny. "Why, 'cause he's old? Most people, even on Rwawan, get married, Clista, and most of them aren't afraid of marriage like you are."

"Stop, please." I reached out and placed my hand on Frall's arm.

He sighed. In fact, for a moment, he almost looked ashamed, but he shrugged and looked away. Then he turned me about and said. "Come on, I want to show you something else."

He led me through the orchard, and we came out into a garden full of vegetables and flowers. "Oh, I don't believe this. Look! This is so pretty! Do you suppose this belongs to Thoric's wife?"

"Maybe they're Thoric's," Frall laughed. "Aren't you being sexist?"

Thoric hadn't seemed the type to bother with colors and fragrances. But then I really didn't know him. I hadn't taken the time to find out who he was. I would have felt guilty if I hadn't been so busy breathing in the garden's aromas.

We walked further into the garden. There were patches of reds, yellows, oranges, and then a plot with mixed colors. I stopped to take a mental picture that I could keep forever. "If this were mine, I'd live right here," I told Frall.

"I knew you'd like it," he said, beaming as if the garden were his doing.

We saw two more robots weeding and tending to the plants. I asked one of them the name of Thoric's wife and found out she was called Breena. I wondered why Breena hadn't met us when we arrived. I thought that it was likely that if she'd been there, I might not have been so apprehensive, but who knows? I guess I'd still have been scared silly after the walls surrounded us and locked us into that chamber.

We came to a large fountain, one perfectly located in the center of the garden's maze of squares and walkways. The water from it shot up halfway to the roof and arched for at least fifteen feet outward. While I stood admiring the falling watery curves and the rainbow lights inside the mist, Frall took off his shoes and socks and urged me to go in.

"No way," I said. "What if Thoric doesn't like us doing that? It's his property, you know. He might get mad."

"Then he can tell us to get out," Frall laughed. "Come on."

The sun was peeking in through the sunroof. It was warm inside the greenhouse's top floor, and the water looked inviting. How could I refuse? In a minute, I had my socks and shoes off and followed Frall into the water. Since the hem of my dress fell to my ankles, I had to hold it up with one hand.

Frall and I started off just wading about, but he splashed me with his feet and hands, and of course, I had to get even. At first, it was just sprinkles, but the water from overhead was shooting down on us, and our returning volleys of splashes got heavier and heavier. Soon, our clothes were good and wet.

Then Frall slipped and fell, pulling me down with him. The reservoir of water around the base of the fountain was only about four feet deep, but with the spray and our tumble, we were swimming in our clothes.

Of course, that was when Thoric found us. "Children," he cried, and we froze in the middle of our play. "Out now. Robot, get towels," Thoric ordered.

One of the robots scooted off in search of a couple of towels, and we, silent with sudden nervousness, climbed out. Our clothes were dripping, forming puddles all about us, but we stood in front of Thoric and waited for the lecture. It surprised us when he only laughed.

"Breena, come and meet Clista. I have found them here in the fountain," Thoric said with no more than a gentle head shake.

Like a drowned rat, my hair soggy and disordered, I was introduced to Thoric's wife, a thirtyish, petite woman with skin like a baby's, hair as curly as a *lopsy pugling*, and eyes that twinkled as she grinned.

"I'm so glad you're finally here," she said, holding out her hand to me. "I would love to hug you, Clista, but under the circumstances, I think I'll wait until you're dry. Is that all right?" she asked, laughing.

"And dear Frall, you look so much happier now than when I first met you. I see what you mean about Clista. She's delightful."

Breena shook my hand with a firm grip and leaned close to kiss my cheek. "Bring them towels, Thoric," she said, stepping back from me as if she expected me to shake water all over her.

"I have sent for some, my dear. Being wet will not hurt them, anyway. It is sufficiently warm in here, you know."

"Well, I suppose so, but this darling girl is starting to turn blue, and her teeth are chattering."

"Ah, here come the towels," Thoric retrieved two soft, plushy blue towels from the robot's outstretched arms. "Here, child," he said as he handed me the first. "Wrap this around yourself before Breena sends you off to bed. Frall, catch," Thoric said, tossing him the other.

The room had been really warm before our dunking, but standing there in wet clothes, not so much. The towel felt delightfully warm. I snuggled happily inside its heavy thickness.

"All right, Breena. She does look chilled. Enough time for introductions. Let's take them to their rooms so they can change," Thoric said.

It was obvious that neither he nor Breena were the least bit upset with us. In fact, Breena actually looked amused to hear Thoric's story about how he'd caught us laughing and splashing at each other in the fountain.

As we walked along, Breena turned to Frall and said, "But I thought you told me you two had been fighting. You must have exaggerated the seriousness of the argument. One can see you've been having a wonderful time together, and it's obvious that Clista adores you."

Frall flushed.

The rooms we'd been assigned were on the fifth floor. Breena ordered Thoric to take Frall to his room. I waved goodbye and watched as they walked farther up the hall.

"Each of you has been given your own room now," Breena told me. "You and the woman shared before, but that was only because you were all a bit mind-fogged by the long sleep. This time, she will be placed in the room next to you. Is that okay with you, Clista? What is the woman's name again, dear?"

"Franca. She's my uncle's fiancé." I glanced around the room. The walls in the new room were starkly white with no decorations. The only furniture was a large bed with a snowy white quilt on top with two chairs nearby. "This is perfect," I told her. "Thank you."

Her eyes discarded my praise. "Thoric told me about your uncle. We must work on his understanding of you and Frall, shall we?"

Breena's eyes were still glancing about the room. "Really, Clista? You like it like this? I would choose color, but that is entirely up to you, of course, my dear."

"Choose color?" I repeated, not understanding.

"But, of course. Pick whatever color you'd like and speak it into the room computer. "Like this: 'Yellow walls, please.'"

Instantly, the room turned the color of mustard. I grimaced. Breena laughed at my expression. "Computer, make that buttercup yellow for the walls. I think that would be better."

I breathed a sigh. What an improvement.

"You may play with it later, Clista. But for now, we need to get you out of those wet clothes. Computer, warm the room."

Breena walked over to the door in the wall, waved her hand, and it opened, unraveling, like the archway outside and the door of the locked chamber, except this one unwrapped itself inwardly, expanding to show a walk-in closet with dresses and jackets in so many colors and styles, I stood there with my mouth a gap.

"You don't like them?" Breena asked disappointedly. "Oh, well. I had a good time selecting things for you, but you can choose new items if you want."

"Oh, no!" I cried out. "I'm just shocked by everything. I can't believe you did this for me. It's so wonderful. I don't know what to say."

"Why, surely you expected nice clothing, dear. You are the queen, after all."

I took a step back. "Please, don't start that. I'm not really a queen. I'm just an Academy student. You see, all this happened because my parents are archeologists and…"

"Clista, tell me about it later. You run into the bathroom now and get yourself a hot shower. Then, when you're warmed up, you can try on the dresses until you find something that pleases you. If the room becomes too warm, simply tell the computer what you want.

"When you're dressed, come downstairs. Thor and I spend most of our time on the second floor. Okay? You and I can talk about anything you want when you're not shivering in wet clothes.

"Oh, and put the dress you're wearing in that hole there, towels and shoes, too. You see it? That's the recycler. It'll dispose of any soiled clothing. Wet or dry — it doesn't matter."

"You throw clothes away?"

"Oh, Clista," Breena laughed. "I keep forgetting this is all so new for you. We'll talk later, dear. Go shower."

"Thank you. Thank you so much for everything. I love the clothes. That was the nicest thing anyone's ever done for me. It's like a whole store inside a closet. Wow!"

She smiled. "I'm relieved you like the dresses, but let me know later — maybe tomorrow, and I will show you how to order new ones. I'm positive I have your size right. I think everything will fit. If not, tell the computer to make it fit. It's that simple."

I was ready to start thanking her all over again, but she hushed me and ordered me for the final time into the shower, her finger pointing me in the direction, her eyes brooking no more delays.

I turned and walked inside, closing the door behind me. I had already acquainted myself with the shower apparatus on the ship. This one worked the same way. Soap, shampoo, and cream rinse were each dispensed at my spoken command.

When I finished lathering my hair and rinsing it, I stepped out into the room's dryer-heater fan. I wrapped myself in one of the fluffy, yellow towels. (I think they'd been matched to Breena's choice of room color.) I half-expected to find the woman waiting for me, but she was gone. Although I liked her very much, I was relieved. I needed time to sort through my thoughts — and all the clothes she'd picked out for me.

There is nothing harder than choosing something to wear when everything is new. All the garments were long, flowing dresses in the same style as the one I'd received on the ship. (It was obvious Flagan Pren didn't like women in pants.) I decided I'd ask if that was a rule. I certainly wasn't looking forward to dresses EVERY day. However, I didn't mind sometimes. These were comfortable ones, anyway.

My hand brushed over the fabric of several gowns, feeling the soft velvet of them. They were like flower petals — no, maybe more like peach skins. I chuckled at the thought and investigated further, admiring the colors and softness.

I finally chose one of the same shade as the grape Smog had shoved into my mouth earlier. I guess you'd call it wine or burgundy, but it was a rich color, and I loved the tone. When I stroked the dress's fabric, my fingers caused ripples, like the surface of sand after raking or the way the carpet looked when a robot vacuum passed over it.

Having decided on that dress, I explored further. Inside the closet, there were drawers containing everything else I needed. I picked out some undergarments from one of the five drawers full of such clothing. As Breena had promised, everything fit.

In the bottom section of the drawer area, I found a cupboard full of shoes. All were the same kind of slipper-like footwear that Franca and I'd been given onboard the ship. Of course, there was a pair in the exact wine color of the dress I'd chosen. Breena had been correct about the size for the shoes, too. How had she figured that out? How had Old Ones known about the shoe size I'd worn on Rwawan?

The last item I needed was a brush for my hair. My long, slightly curly hair was still damp, but it was drying quickly. Already, I could feel the snarls forming as if there was such a thing as coiling magic. My hair had always had a tendency to twist itself into knots if I didn't attack it right away. Unfortunately, although I checked every drawer in the closet, a brush was the one thing I couldn't find.

I slipped my dress on and left my personal dressing room closet. Then I focused on raking my fingers through the damp mass, attempting to prevent what I feared, but the tangles were already setting. I walked over to the bed, sat down, and, with a huge sigh, began separating it strand by strand.

"Come in," I called out when someone knocked. It was my uncle and Franca.

"Oh, we're so glad you're here," Franca said. "We were looking for you."

"I've been sitting here at least twenty minutes trying to get the tangles out of my mop. I can't find a brush, and . . ."

"Let me see what I can do," Franca offered. "You look lovely, by the way. That dress is just the right color for you."

Franca took a seat on the bed beside me.

My uncle walked to the other side of the room. "She's right," he agreed, giving me a quick inspection. He dropped into a chair and perched his feet on a small, round table that he'd scooted in front of him.

"Breena showed me my room," Franca said, "but I haven't had time to change yet. I will before dinner, though. Did you know that Thoric and his wife invited us to dine with them?

"Just what we wanted," my uncle said sarcastically. "No problems with Frall, Clista?"

I hadn't known about the dinner invitation. That meant that I couldn't put off the chat I needed to have with my uncle. He needed to stop attacking Frall. It wasn't fair, especially when he and Smog kept ganging up on Frall. My boyfriend didn't deserve their constant barbs. Besides, if Uncle Cristoff kept interfering with Thoric's wishes, who knew what the *Old Ones* would do?

"Uncle Cristoff, I need to talk with you. Please."

"You want me to leave, Clista?" Franca asked.

I patted her leg. "No. Stay, please. I want you here."

My uncle shifted, dropping his feet to the floor. His eyes turned surly, and he crossed his arms. I could see already that he wasn't going to be receptive to my request.

"Uncle Cristoff, you know that Frall asked me to marry him, and . . ."

"And there's no way I'm going to allow that. You're a minor still . . ."

"Cristoff, let her speak, please," Franca interrupted.

He shook his head. "You don't understand, sweetie. They're just kids, the two of them. It would be a mistake . . ."

"I understand that Clista told you she wanted to TALK to you."

My uncle understood what she was saying. I saw it in his face. He sighed and stared down at the floor. "I'm sorry, Clista. It's just that I want what's best for you. You understand that, don't you?"

"Yes," I said, still waiting for my chance to speak.

He sighed again, long and lingeringly. "Okay. What did you want to say, sweetie? I'm sorry I butted in."

I was quiet another moment, waiting to see if he meant it, if he really was ready to let me talk. His eyes lifted from staring at the carpet, and I saw that he wanted to give me his attention and would try to be patient and give me my chance. (Fingers crossed.)

I started again, explaining a bit long-windedly, perhaps, but I just had to make him see that he couldn't keep picking on Frall all the time. "I just brought that up about Frall," I said, "because I want you to understand that Frall loves me. And the truth is that I love him, too."

"Oh, Clista, I know you think you do, but . . ."

My uncle was off and running again. I watched as his eyes grew firm with fervor. He wasn't looking at me. He apparently lacked the ability to listen.

"Cristoff," Franca said gently, shaking his arm slightly to interrupt his lecture.

My uncle stopped and turned to look at her. I saw when it hit him that he'd done it again. He raised his hands vertically in a gesture that meant he'd stop — another promise that he'd try to be silent.

"I'm sorry, Clista. You were saying?" he asked, nodding for me to go on.

I darted a quick smile at Franca. I was glad she was in the room. "It's just I don't want you to attack Frall all the time."

"You don't want what?" His eyes widened. "That's what this is all about? You just want me to . . ."

"Please. It's hard enough with all the pressure on us from Thoric, and I don't know what's going on with the other *Old Ones*. I think Frall's just as confused as I am about it. I don't even know if he really, really wants to marry me. I mean, it might just be . . ."

"No, you're wrong there, Clista," my uncle barged again. "I've spoken to him, remember? The kid's head over heels in love with you. That has nothing to do with anything Thoric, and the *Old Ones* are ordering him to do. Believe me, I've seen that look in a man's eyes before."

"I thought you were going to let Clista talk," Franca said with a smile so warm and full of love, I was reminded of exactly how she felt about my uncle.

This time, I was the one who sighed. It was so easy for them. They were ready for it, old and settled down. They were the ones who should be getting married. I wondered if Thoric knew someone who could do the deed. Did they have marriage officials on Flagan Pren?

My uncle paused and took in Franca's reproach. "My charming school psychologist. I've done it again, haven't I? Sorry, Clista." He chuckled. "I'm new at this, you know. I guess your parents did it a lot better, didn't they?"

He looked defeated, as if he didn't stand a chance with the competition. I almost laughed. He was actually doing a great job of being a parent — at least of being a normal one.

"My parents probably wouldn't have even noticed," I said. "They were pretty caught up in artifacts and stuff."

"Clista, you don't mean that," Franca said, rubbing my back and giving me a quick hug.

"I don't mean that I don't love them. It's just they don't . . . I mean, they didn't . . ."

"You can still use present tense, Clista. They'll always be alive in your heart. Go on," Franca urged.

I sighed. I knew Franca meant well, but my parents were gone. I had to accept that.

I started again, trying to explain what I meant. "I know it sounds awful — what I said. But it's true. I loved my parents, of course, and I know they loved me, too, but I want Uncle Cristoff to understand that he doesn't have to worry about competing with them. I mean it. My parents never saw much of what was going on with me, not ever."

"You mean you and Frall were a couple before all this happened?" my uncle asked. I suppose he was wondering how close Frall and I had been.

"No," I said, shaking my head. "I mean, Frall and I were friends, and we met every day before school, but . . ."

"And your parents didn't know about that?" Franca asked.

"No. Not until the day, Mr. Grenan — you know what day . . . Well, I got sick because of that, and Frall saw me as Dad was taking me home. I fainted, you see, and Frall carried me out to the car. That's how Dad found out about Frall.

"The next day, Frall stopped by the house. He ate dinner with us, and the *Old Ones* demanded that I show Frall the artifacts. That's when we found out that he was receptive, too. Anyway, that's the first time that my parents really got to know him."

"And?" my uncle prodded.

"Well, they liked him. They were going to let him work for them in the caverns. But I'm glad he didn't. If he had, Frall would be dead, too."

"Or your parents might still be alive, Clista," Frall said from the open doorway.

"Don't even think that," Franca said, bolting up to give him some back rubbing support. "It doesn't do any good to make assumptions. What's done is done. It can't be changed."

Frall sighed. He allowed her comforting a moment, then stepped away. I could tell it had made him uncomfortable.

"If I'd been there in the caverns, don't you see the *Old Ones*, the ones from Flagan Pren — they would have protected me. By doing that, they would have saved Clista's parents," Frall explained.

"Hold it," my uncle interrupted. "Your presence wouldn't have meant that. If it's true about your being protected, maybe you would have escaped, but that doesn't mean it would have changed what happened to Clista's parents."

"Do you feel guilty for not being there?" Franca asked in her soft, gentle voice, the one I instinctively recognized as her therapy voice.

"Not being where?" Smog asked, intruding at the wrong time.

The moment had passed. Frall stiffened up, shook his head, and said, "Never mind. It doesn't matter. What happened can't be undone."

"Are we all ready for dinner? I see Clista is. You look nice," Smog said, his eyes smiling his appreciation of my dress.

Franca rose. "Oh, dear. If it's that late, I'd better start primping." She laughed good-naturedly and headed for the door.

Uncle Cristoff took that as his cue to begin telling Franca how beautiful she was just the way she looked. Of course, she started arguing about it. They left, both disputing whether she was pretty or not and whether the dress she was wearing made her look fat. Presumably, they headed off to change for dinner.

Smog, walking the rest of the way into my room, sat down in the chair where my uncle had sat, and made himself as comfortable as if he'd been invited into my room.

"Take a chair, Frall," Smog ordered. "Let's keep Clista from getting bored while Franca and Cristoff get ready." Smog gave a quick yawn and stretched out his legs, leaning back like he was considering getting forty winks. He repositioned the little table Uncle Cristoff had used as a footrest and then looked up, his eyes darting from me to Frall, who still stood in the doorway.

Then Smog closed his eyes and said, "Frall, why don't you tell us what you learned from Thoric."

I wanted to hear that, but there were other priorities at the moment. "Before you start that, did either of you happen to find a comb or brush in your room? I'm desperate."

"Is that what you're trying to do?" Frall grinned, eyeing my fingers all caught up in hair tangles. "Actually, I rather like your hair like that. You're a wild woman!" he laughed.

I sputtered a moment, not knowing whether to argue or laugh.

"But if you insist, I'll go get the brush," Frall told me, winking at me as he entered my bedroom and headed into the bathroom. He returned almost instantly, carrying a small appliance that looked more like a backscratcher than a hairbrush.

"That? I'm supposed to use that on my hair?" I said, examining the object with considerable disbelief.

Frall laughed. "Let me show you. It won't hurt. I promise."

Frall came closer and sat down on the bed beside me. Smog's eyes opened. He pulled his legs off the table and sat up to watch.

"Help, Smog," I cried out, half-kidding, half-serious, worried about Frall making my tangles worse.

Big help Smog was. He just shrugged as Frall lowered the *Old One's* brush down onto my head. The part with all the tentacles was placed down into my scalp. It immediately went into action, massaging and smoothing, working its finger-like digits around and around as it slipped down the length of each strand of hair.

Frall held it in place until it beeped, and then he lifted up the appliance and stood back.

"It's done? What does it look like? Is it awful?"

"You look incredible," Frall said as he left to return the brush to its compartment in the bathroom.

"Thanks," I called after him. "Does it really look okay, Smog? Really?" I touched my hair gingerly. It felt soft and manageable. But what did it look like?

Smog nodded as if that answered my question, but he was smiling. Still, I wasn't sure if I trusted the guys. I half believed they might be playing a trick on me. I jumped up and ran into the closet, where there was a full-length mirror.

"Wow!" I yelled out. "I absolutely love that thing. My hair looks great! No tangles, no frizz. It's got body! Wow!"

"I think she likes it," Smog told Frall, chuckling.

I wished I had a second mirror so I could get a good luck at what the brush had done in the back, but I heard Franca and my uncle talking out in the hallway. They returned to stand in the doorway, dressed and ready to go.

"You're not going to believe this," I told Franca, stepping out of the closet and waving it shut behind me so it could cascade into itself and go back to being smaller. "But there is a really neat brush in the bathroom that massages and fixes hair."

Franca smiled. Her eyes slid over my much-improved hairstyle. "I'm impressed. It was badly knotted before. Show me this marvel, please," she urged, patting her own hair.

"Not now, ladies," Uncle Cristoff said. "Dinner is calling."

"Amazing. I never heard it say a word," Smog joked, sticking his finger in his ear and pretending to clean out his earwax.

"You're so funny, Smog," I laughed. "I don't know about dinner talking, but I sure know my stomach is."

"Mine, too," Frall laughed, sliding his arm around my shoulder.

I saw my uncle frown and start to speak. I gave him a look. He sighed heartedly, then turning, put his own arm around Franca, and the two of them led off.

It was really nice to have everyone friendly and happy again. I pecked Frall with a quick kiss just to let him know I was content to be with him. Smog, following from the rear, popped off. "Hey, none of that, you two. You've got a chaperone behind you."

I giggled, knowing Smog had whispered so my uncle couldn't hear. I sure hoped everyone remained on good behavior for the rest of the evening. I guess that depended a lot on Thoric. If he and Breena didn't mention marriage, dinner would go a whole lot better.

We all stepped into the same elevator. Apparently, its glass cage expanded or contracted depending on the number of people inside it, just one more facet of the futuristic technology surrounding us. Dwelling on the advancements of the *Old Ones* could really give a person an inferiority complex.

There was no sense of movement as the elevator lowered. We could see through the clear windows, outside the building or inside on a lower level, depending on which glass wall we were looking out. I raved about the wonderful views, but Franca, poor lady, turned pale and asked me not to discuss them. She kept her eyes shut and refused to look.

When we reached the second floor, a force field prevented us from stepping out.

"Just wait a minute," Frall said. Thoric will release us in a minute."

He was correct. Almost at once, the misty blue light of the field stopped flowing across the open door, and we were free to enter the huge chamber.

"Welcome," said Breena, walking toward us. "Did you all rest pleasantly? I see the clothes fit each of you quite well. I'm so pleased. Sometimes the scan gets something wrong, but this time it did a great job."

We had no idea what scan she was talking about, but none of us interrupted.

"Clista, you look lovely. You have such good posture. Wonderful skin, too. Isn't she beautiful, Frall?"

Frall flushed. "I think so, Breena. I told her that, too."

"And, Clista, what do you think of this mode of dress? Is it appealing to you?"

"Uh, Breena," my uncle injected, rather rudely, I thought. "Can you tell us if we'll be meeting others?"

"Others?" Breena asked, stopping in puzzlement. "Oh, you mean more *Old Ones*, as you call us?" She laughed.

"Don't be embarrassed, child," she scolded me as I groaned. "I know you don't refer to us that way based on our ages," she said, giving me a pat and laughing again.

"To answer your question, Cristoff, I'd have to check to see if any of the citizens of Flagan Pren are here on the planet at this moment. They rarely stay in one place. Thor and I are quite an exception.

"Are you wanting to meet others? We can issue an invitation for them."

"But we assumed that this was a planet full of people," my uncle said.

"There are perhaps sixty dwellings on it, I suppose," Breena continued, speaking as she once more led us forward. "But, as I said, most people flit about, enjoying the sights of traveling the galaxy."

"There were settlements on Rwawan. Didn't you once live there?" Frall asked.

"Clever boy. He is such a delight, Clista. Don't you think so?"

I nodded but didn't say anything more, feeling the eyes of my uncle scowling again.

Breena laughed again with a voice that tickled, vibrating the hairs on my skin. Frall joined in, too, smiling into my eyes, letting me know he'd shared my thoughts.

"First of all, as I think you suspect, Frall, Rwawan is our ancestral home."

"It is?" I cried out. "Why did you leave there, then?"

We'd arrived at the table. Breena gestured for us to sit. "Thoric will be here in a minute," she said. "It is his customary time to commune with the others. I'm sure he would have been done by now if they hadn't had so many questions about you. Your appearance on Flagan Pren has quite excited everyone."

"Shouldn't that send them returning here in droves?" Smog asked, puzzled by the continuous mysteries we kept finding.

"Ah, poor lonely man. We have sent for a partner for you. It is not good to be unmated. She will arrive soon."

Smog blinked a couple of times and then made a show of clearing his voice as he tried to figure out what to say. "Uh, uh . . . is that something we should discuss in front of the kids?" he asked, his face turning the deep, red color of an alarm switch.

"But that's great," I said. "Smog asked if he could bring a girlfriend just before we left Rwawan."

"Yeah, but I didn't really think they'd pick out one for me," he said, shaking his head at me, his face blooming even redder.

"We were impressed with the elevator, Breena. The view was quite panoramic," my uncle said, changing the conversation.

"Oh, did you enjoy that? Then we must take you up on the roof. You can see even farther from up there," Breena said, obviously not minding that we'd jumped subjects again.

While Breena was talking about the landscape that could be seen from the roof and telling us that one of the others' dwellings was visible from the south side, Thoric joined us.

"So what have I missed?" he asked as he stopped to plant a brief kiss on his wife's cheek.

Smog muttered something no one caught.

"What did you say?" Frall asked.

Smog was staring at Thoric, a look on his face as if he'd just seen a ghost. "I said, this is too weird. We've gone up into space and flown almost a year to arrive on a planet where nobody lives except for one couple that acts exactly like a human couple, even down to the kiss on the cheek when the husband arrives home. Something smells rotten. None of this makes sense."

I hadn't thought of it that way, but Smog was right. Everything did seem an awful lot like Rwawan.

"I'm sure that all people react more or less the same," Franca said, smiling an apology at Thoric and his wife.

"Yeah, they do," said my uncle. "Unless they're alien. Why didn't I see it? I've been so involved with watching Clista and . . ."

The two men stood up as if making such a speech had ejected them from their chairs. The rest of us remained seated, but Thoric, frozen by the side of his wife, his hands on her shoulders, stared at me, watching my eyes as if I'd been the one who'd just spoken.

"Come, do not be so suspicious," he said. "Sit down, everyone. Let us eat a pleasant meal together."

Something clicked inside me then. The puzzle pieces fit. "This is all an illusion, isn't it?" I asked, interrupting everyone.

Chapter Five

There were many moments in math class when equations and shapes, theories, and teacher input used to swim about inside my mind, oddly disconnected, lacking meaning and substance. I'd tried to force them into a line-up, matching them like puzzle pieces, but most often, they didn't correspond, didn't even belong in the puzzle.

It was like there was more than one puzzle — in fact, half a dozen different ones, all mixed up together. I kept working at sorting them, trying to find the edges. My friend, Bista, had an eye for seeing it. He'd handed them to me. With his help, sometimes, I could make the connections. And then there were days, days when the puzzle pieces realigned as if they were dancing into shape, arranging themselves, as if by magic. Then, I could see the whole.

It was like that at the moment when Smog said that about things being too much the same. All the puzzle pieces started dancing about, kicking up their legs like in a Cancan line, and then they fell, did a quick scramble, and reformatted. Although I didn't exactly "see" the illusion, I sensed it, and I knew.

Strangely, no one commented. They grew silent. Thoric stared into my eyes as if he was reading my thoughts. Perhaps he was. "Does a little fantasy hurt?" he asked. "Doesn't it make you feel more at ease?"

"No," I said, shaking my head. "I want to know the truth. I don't want to believe in lies."

"But maybe you still need the illusions, Clista. Do you wish to take away the substance?"

"I don't understand you. What do you mean by 'take away the substance'?

"Frall, you knew, didn't you?" I asked. "Yet you never told me."

I shot a glance at the others. Thoric had frozen them into stasis. My uncle stood, glaring at Thoric, his finger outstretched as if he were ready to deliver a lecture. Smog was right beside him, in a defensive position, his feet rigid and ready to pounce, looking as if he could tackle the enemy and bring him down, something he'd probably done many times before. Franca still sat with her usual gentle, quiet smile. She was looking at Thoric trustingly.

"You rejected what he offered, Clista," Frall said. "Thoric's only trying to please you with all this. As I do, in not pushing you before you're ready."

"Then everything's fake?" I asked, looking around the room. "But what part is real?"

Thoric moved away from his wife. She was frozen, too. Something about that bothered me. I stared at her. "Wait! You said stasis doesn't work on the *Old Ones*."

Thoric laughed. "There are very few of us left, Clista. My wife does not know it, but she is not one of us. She is of Terran origin, like the others."

"She's not an *Old One*? But she's so comfortable with everything…Oh, I don't understand. What *is* real?"

"All of us, my dear, but nothing otherwise. Everything is created from our desires. Choose a new fabric, weave a design. . . it is what we do. We do not need buildings and possessions now. Most of us have become spirit dwellers, flitting among worlds of solidity, joining for a time, then dissolving back into our true nature."

I stood up and walked toward Thoric. Frall reached out to grab my hand as I passed by, but I kept going, dodging his touch. Then, when I was near the *Old One*, I reached out to feel his arm. He felt solid. I'd half expected my hand to pass through him, but he had skin and bone and substance.

Thoric laughed. "Your eyes are such a barometer, my dear. What is it you want now? Flash them and show me. Do you want the truth or the illusion, dear Clista?

"How much can she take, Frall? Guide me in this."

"I don't understand," I said once more, glancing over at Frall.

"Give her the whole. Let her see," Frall replied. "She has asked for it."

"Touch me again," Thoric said. "Feel my nature, then."

I didn't want to after he said that. I was afraid. Sometimes, when you unlock a door, you wish that you hadn't. Yet, I'd stepped forward, leaving behind the comfortable. I lifted my hand, and my fingers reached out. This time, there was nothing there, just a shimmer of light and fading color.

I screamed and backed away. Frall rushed to me. "Accept them, Clista," he said, holding me in his arms. "Accept."

Frall was real. I felt his arms and his chest, the beating of his heart. I pressed my head against the strength in his heavy chest. "No. This is a nightmare. I'm dreaming, Frall. Wake me, please."

He picked me up and carried me, walking toward the huge stone throne. It was then I realized that we were in the huge chamber where we'd arrived that first day.

"No," I cried out. "Please, Frall. Don't do this," I begged, for he was taking me to the throne. Instinctively, I'd been afraid to go there. Something warned me not to sit on it, but Frall wasn't listening to my pleas to put me down. He kept walking until he stood in front of the bigger of the two thrones.

"I love you," Frall said. "But the time has come. Don't fear the *Old Ones* or me, Clista. We will rule the stars. It is our birthright. You must learn to accept that."

So saying, he placed me down on the stone-cold throne, and I was caught.

Thoric had reverted to substance again. I saw him standing on the floor down below me, watching as the throne caught me up and whirled me into the chaos. I closed my eyes against the dizzying, churning lights circling round and round my body. The throne was a spinning top, twisting and stopping to revolve once more, each time in the opposite direction and spiraling downward. My heart raced. My blood pumped in frenzied spurts of panic.

"Clista, I'm here beside you," Frall said.

I turned my head to look at him. He was sitting on the other throne, not scared at all, but relaxed and calm. I shut my eyes and crouched farther down in the chair, cementing myself to the armrests with my terror-stricken grip.

"Why?" I asked Frall. "Why did you do this to me?"

"Because you keep advancing and then retreating. Because you are refusing to see. I know that you won't accept any of this otherwise."

We were still spinning in our hurricane of circles. It shot us downward into a deep and dark well of black. "Where is this taking us, Frall, and why?"

He laughed. He was that relaxed. I almost couldn't forgive him for that. My horror at the ride was smothering me.

My ears felt sudden pressure changes. My stomach hurt from clinching it, and my hands were stiff and cramped from the cold, hard stone I was clutching. I'd already clenched my teeth 'til they ached. "I don't like this," I told Frall in the understatement of the century.

"I know," he said. "I feel your every thought. Relax, Clista. You know I wouldn't hurt you. We will reach the bottom now, and you will understand soon. You will forgive me then."

"Frall, I'm too frightened. I don't want to see it. I just want to go home."

"You have the *Old Ones'* blood in your veins, Clista. Focus on that. It will give you courage."

I swallowed hard. The throne was racing us through galaxies, speeding me further into terror. "What if we can't breathe? What if we . . ."

"I will always keep you from harm. I am your mate, Clista. Feel that. Know it. There is no danger."

The spinning escalated. I shut my eyes tight enough to see colors and stars. I could barely breathe, but I knew that was due only to my fear and not to any lack of oxygen.

"Frall, please, make it stop. Please."

"We are almost there. Take my hand. I will give you strength."

There was no way that we could touch with the distance between our thrones, but obediently, I stretched out my hand, and Frall took it. His warmth and the feel of his closeness settled my stomach. I breathed in my first full breath since sitting on the throne. My racing heart slowed slightly.

"Thank you, Frall, but I wish I'd never asked about the illusion. I want the lies back. I want normality."

"No. We can't go back. We've come too far."

"My uncle and Franca and Smog, what about them?"

I felt Frall's smile. I knew his eyes were on me. "I adore you, Clista. Did I tell you that yet today? Even in your fear, you are still thinking about the people you love. Be that constant with me. Love me, Clista. Love me always, no matter what I do to help you go forward."

"Oh, Frall. I do love you. Even here, even after what you did, I still love you. But where are they — I mean, Smog, Franca, and my uncle? Are they still all right?"

"Of course. They're where we left them, Clista, at the table. They're in stasis, remember? They'll remain that way until we return. No one can hurt them. The *Old Ones* protect them."

"They didn't protect our parents, Frall."

Frall sighed. "Clista, you can be so difficult at times."

"But it's true."

"The *Old Ones* have promised to protect your friends, Clista. They never go back on a promise."

The chair I was sitting on was galloping through space, and I was worrying about people who were safely on the ground. It didn't make a lot of sense, I guess, but it relieved some of my anxiety. "Thank you, Frall," I told him. "Knowing that they're safe, I'll try to be braver. But where is this taking us, Frall? Are we going to see the *Old Ones?*"

"Yes. The pretense has been cast aside."

My chair began to shake. Its vibration rattled my teeth and made my buttocks sore. "Are we crashing?" I cried out, my fear taking over once more, although I'd just spoken about being brave.

Frall laughed. "Clista, I keep telling you. You're safe."

The touchdown was gentle, a brief slap on the pavement, and then silence. Frall stood up and came toward me. "Open your eyes, Clista," he said. "We're here."

I would have flung myself into his arms, but the shock of my sight kept me gripping the side of the throne. I thought at first that we were underwater, for it looked like water, except there were small, silver particles drifting all about us, like a steady fall of rain, and I was breathing. "What is all this? What's happening?"

Frall gently disengaged my clenched fingers and pulled me up. His arms enclosed me, the feeling so reassuring that, for a moment, I accepted its stability. Frall didn't push me away, either. He held me, his hands soothing me, playing with the long fall of curls down my back.

"We're in the center," he whispered. "Only a little farther now. They will soon greet you, Clista, and then, we will sit on another throne, a larger one than these, and we will occupy it together."

"No more thrones," I begged. "Not again."

He chuckled, and then his lips seized mine. He kissed me lightly, but I felt his touch inside my mind. When he freed me, I was changed. It was like there was an echo of his kiss chasing all the thoughts around and around inside my brain. I threw my hands over my ears, but he took them away and brought them back down to my side.

"Let it settle in," he said, "I'm simply chasing away your fear."

Silver was still raining down, filling our mouths and the spaces all around us, yet it was not that which frightened me most. It was Frall's certainty and his invasion into my mind.

"How do you know how to do the things you do?" I asked. My voice sounded deeper in the silvery water, yet the huge fear inside me had no sound.

Frall's voice boomed just like the *Old Ones'* voices. That thought worried me, too. "I dreamed all this, Clista. The *Old Ones* gave me the information first so I could guide you through it. But your trust in me has not been strong enough. That's why I've had to force you. You must trust me, Clista. To hurt you would be to hurt myself. Don't you see that?"

Things were too weird for trust. "I'm trying, Frall, but none of this fits together. There's no scaffold, no continuity. And you're frightening me worse because you're too much like the *Old Ones*. . .

"Tell me. Am I dreaming this?" I asked.

Frall shook his head. The silver rain painted sparkles in his hair. They clung and glowed. Frall's eyes were glowing, too. "I am only Frall, Clista, the same boy you came to know at school."

I looked down, trying to see my feet, but silvery speckles covered them. Were we drowning in glitter? Would it wash us away into nothingness?

"Look at me, Clista. Why do you still doubt me? All this time, I've been at your side. I've never done anything that could hurt you. Don't you see that? I've moved so slowly with you. Sometimes, it's so slow that it makes me crazy. I want you to leap into the transition, but I hold back, waiting. Don't you see how hard that is?

"You *have* to trust me, Clista. You must."

Frall kept changing, evolving, I suppose, going deeper and deeper into the *Old Ones'* ways. *Could* I trust him? I wanted to. I needed to, but I was so uncertain.

He raised my chin, so I'd meet his eyes. "You asked me if this is a dream, Clista. No, it's real. It's the world of the *Old Ones*, your world. You do understand they're our people, right?"

I shook my head.

"We have their blood, Clista. We're their children."

"You and Thoric keep saying things like that, Frall," I said. "But it doesn't make any sense. It confuses me. You know who my parents are . . . were. Why would you say that about the *Old Ones*?"

"Your mother was of their lineage, and her mother, and her mother's mother. And my father, grandfather, and great grandfather. It goes back many, many years for both of us, Clista, but once a long time ago, an *Old One* sired a child on Rwawan and, in your case, bore one. Then, many centuries later, the two of us are the result. Now they've brought us together."

"The *Old Ones* told you that? But it can't be true, Frall. Aliens can't mix with other races. We learned that in school."

"Our teachers didn't know the truth. What they told us was false, Clista. Smog could sire a child with the *Old Ones* if he desired it. Or your uncle. They can if they want. Our textbooks are wrong. The root of all races — even alien races is close enough to be the same."

The silver speckled liquid was stirring more restlessly about us. My eyes watched it, worried by its increasing thickness. Frall turned. "They're coming," he told me. "Trust in me, Clista. Don't be afraid of their changing shapes. They can form into whatever you like. They don't want to frighten you. But I have asked them to show you their reality."

If Frall were really able to read my mind, he should know I wasn't ready for this. I was too terrified. The silver water blew, storm-like all about us. made me feel like I'd been captured and was now imprisoned in an airless container, an empty fish tank of sparkles.

"Please, please get me out of here," I whispered, with a fresh onslaught of panic.

Frall tightened his arms and whispered. "They come. You'll see."

I saw them then. The *Old Ones* were ghost-like, mere shadows with a cream-like whiteness. Their bodies circled about us in long and slender flowing curtains of substance, and yet not, for the *Old Ones* wavered in their unstableness. I wanted to scream, but I didn't dare open my mouth.

The gray drops of silver were all about me, and the wraithlike beings kept circling and disappearing, fading in and out, reappearing as I stared. Only Frall's arm holding me close to him kept me from running off as the gray liquid dropped, and my hysteria threatened to send me into the catatonia of the insane.

"Well, well, welcome, queeeeeeen," the *Old Ones* sang, with articulations high-pitched and musical. "Were are happyyyyy, happyyyyyyy, happyyyyy, you have fine, fine, finally come, come, come to us. Weeeeee will train, train, train, you, you know, know, know. . ."

"I want to go home. I don't belong here."

"You are stiiiiiill afraid, fraid, fraid. We are sorryyyyyy, sorryyyyyy, sorryyyyy. We wiiiiill nev, nev, never hurt, hurt, hurt you. You are one, one, one, of us, us, us. Take her home, home, home, Frall, Frall, Frall. Let her, her, her come back, back, back ano, ano, another tiiiiime, tiiime, tiiime. Weeeeee wiiiiiill wait, wait, wait, for you, you, you, our queeeeeeen, queen, queen. . . Weeeeee will wa. .ait, wa . . . ait, w . . . ait."

"Stop! What do you want with me? Why did you urge Frall to bring me here?"

"Another tiiiime, tiiiime, tiiiime, we wiiiill talk, alk, alk. You, you, you wiiiill come, come, come, ba . . ck, ck, ck. Then, then, then, you wiiiiiill not, not, not fear, fear, fear us, s . . .s . . . s, so great, t, t, ly. A no, no, nother tiiiiiime, tiiiime, tiiiime, Cliiiiiista. Come, come, come back, ack, ack…."

The specters were swimming faster and faster around me, circling dizzily. "Frall," I called out, terrified by their speed and the way they were entrapping me.

"It's all right. They won't hurt you," he told me again, his arms holding me more tightly against his chest.

I closed my eyes and tried not to listen to the whirlwind sounds of the Old Ones' passing. In a few minutes, the echoes of their chants faded like the wind after a stormy day. And then the noise was gone, and Frall and I were alone.

"Are you all right, my love? They're gone, you know. But still, I hear your heart racing. Breathe deeply. It's over, Clista. They won't come again, not this time."

I was shaking from the delayed reaction. My breath came in sobs and quick spasms of breath. "Can we go back now?" I asked, barely able to voice my hope.

"Yes. But we have to return to the thrones. We have to ride them back to Flagan Pren."

"To the place of illusion?" I asked, walking through the silver, my unheld hand swimming through it, my legs treading its thicker mass.

"It is the halfway place, Clista, the transition between your readiness and what you are not ready to accept."

"I don't understand. Why are you able to acknowledge all this? You resisted once. You refused to even admit that the *Old Ones* were still alive."

"Yes. I went through disbelief at first, and then I accepted the rest. It was meant to be that way, Clista. I am your guide. I told you that."

We reached the huge stone relics they called thrones. I halted before mine. I wanted to go back, but I was afraid. "Do I have to sit on it alone, Frall? Can't we sit together?"

Frall smiled. "I wish we could, Clista, but we need to take them both back. I'll hold your hand again. You will not be alone. Next time, maybe we can try just one. I'll ask Thoric."

"Next time?" I repeated, shuddering at the thought. "I'm not coming back here, not ever."

Frall didn't say anything. He just gently pushed me back onto my throne. Once more I felt fastened in, as if something had fastened my body to the throne.

Frall went to his own seat. "Ready?" he called out. "Take my hand and close your eyes. We'll return soon to your uncle and the others. Okay?"

Before I could answer, the throne lifted up and sped us away into a mad rush of swirling air. I felt no changes of pressure, no lift-off, no ears popping, but when I opened my eyes, I saw lights flickering, scenes passing by, and the stars, moons, and planets of a hundred solar systems.

Frall's hand played with my fingers and squeezed my hand. I closed my eyes. The vibration of the thrones and the faint feel of an inexplicable wind on my face were the only signs that we were passing through time and space.

The giant chairs returned us gently to the throne room. Thoric stood there waiting for us. He took a step forward and then stopped. "Is she all right?" he asked Frall.

I felt Frall's eyes on me. I turned to look at him. He smiled. "Clista was too frightened to achieve Mistah, but she met them. She saw. They circled her and welcomed her into the fold."

"Good. It is a beginning. Come, Clista. Rise up. It is time for that dinner I promised you."

I stood, but my legs were too shaky to walk. I clung to the throne, holding onto its armrests. Frall walked over to me. He picked me up and carried me. I placed my head on his shoulder and wept.

"I'm such a coward, Frall. I'm sorry. You told me there wasn't any danger. Why didn't I believe you?"

"It doesn't matter, Clista. The *Old Ones* were right. Next time, you won't be as frightened." Frall kissed me. I grasped his body, clinging as I had a moment before to the throne.

"Are you okay now?" Frall asked, echoing Thoric's question.

How could I do anything but nod? I hadn't been hurt, but I felt weak, both physically and mentally. This clinginess wasn't like me. I sighed, asked to be set down, and then walked to the table — almost as if nothing out of the ordinary had happened.

Chapter Six

"The others will not know you left," Thoric told us as we returned. I glanced about and saw that my uncle, Smog, Franca, and Thoric's wife were still frozen statues, awaiting a counterspell.

"So what happens now?" I asked. "Do we go back to the illusion?"

Thoric laughed. "The chimera of fantasy lives in all of us. Why should we not savor its daydream for a while when we like?"

"But it's all fake. You told me so."

"Is it?" he questioned. "My wife delights me, yet she cannot exist in Herva, so I spend most of my days here with her. You can, also, if that is your preference."

"What about you, Frall?" I asked, turning to look at him. "You liked it there in that place, didn't you? Do you want to go back?"

"I will go when you are ready, Clista. We will go there together."

Thoric smiled. "And until then, we will live in Flagan Pren. Are we ready for our meal?"

They were both waiting for my answer. I searched their eyes a moment, then glanced again at the unmoving sculptures that were our family. I nodded, but I wondered how one could flit between such worlds? How could you pretend to live in reality, knowing that it was all a sham?

"It would be wiser not to discuss Herva and the things you have seen, Clista," Thoric warned me. "Humans are fragile-minded. They would probably not believe you, but it might give energy to their doubt. Sometimes, humans fray at the edges when their reality is disputed. Do you understand?"

I nodded. My reality was in dispute, too. Maybe I was fraying.

"Well, isn't this lovely," Thoric said, sitting down. I stared. My mouth turned goldfish-like as my uncle and Smog slipped back into their chairs. Breena was requesting that Frall pass the mashed potatoes. She, herself, was starting a platter of something that looked like soy-chicken cutlets. Thoric passed the gravy. Smog picked up the biscuits and helped himself to two. Then he grabbed up some jam and ladled it onto his plate.

"Wake up, Clista," Franca said gently. "Frall's trying to hand you some vegetables."

I reached out and took the platter, put some vegetables on my plate, and handed them on, but the shock of the change was overwhelming me.

"I can't do this," I whispered to Frall.

Pretend, he thought back at me. *Pretend a moment, and you'll slide right into it.*

"Hey, will you two lovebirds stop dallying and start handing things around?" Smog jeered.

"Oh, leave them alone," Franca ordered. "It's entirely natural for them to cling together. They're the only teenagers here, in case you didn't notice."

"Yes, that certainly is true," Breena said, laughing. "Isn't that sad? Maybe we should send for some others to make the kids feel comfortable, Thoric."

He laughed and shook his head. "Sweet, Breena, I'm afraid I don't think that's a good idea. Besides, I have a feeling that wouldn't change anything. Frall and Clista would still be whispering together. Couples do that."

Frall, apparently sensing that I was at the end of my tolerance, sweetly changed the subject. "How many trees do you have in your orchard, Thoric?"

"Thanks," I whispered as the conversation shifted away from us and into safer territory. Frall handed me the potatoes and watched as I took a small spoonful. "Dieting?" he asked.

"None of your business. Then I laughed because Frall was right. Pretending normalcy did encourage it.

The rest of the dinner passed smoothly. Thoric and Breena stayed clear of the dangerous areas, my uncle remained on good behavior, and Smog was too busy stuffing his face to come up with any more of his side cracks.

It wasn't until everyone had pushed their plates away from their stomachs that Smog started up again. "So, is this what aliens eat?" he asked casually.

Thoric laughed. "We have served you a Terran meal. Breena is partial to that kind of cooking. She thought it might please you, as well."

"It was a lovely dinner, Breena. Thank you so much," Franca said, smiling at everyone. "I can't say when I've enjoyed a meal more than this one."

"Yeah, it sure beats what we used to eat in the forest, doesn't it?" Frall said.

"Oh, but tell me about that," Breena said. "What did you eat in the forest?"

"Fish, every morning," I said, "and coffee made of roots and nuts."

"Which you raved about," Frall laughed.

"Well, it was better than that stuff we drank before Smog found the nuts. Root water made a very weak and bitter coffee, believe me."

"But how adventurous," Breena said. "Imagine living off the forest. Did you hunt for wild animals, too?"

My uncle jumped in with his two cents then. "We roasted our catch on wooden poles. We even made jerky so we could carry it with us."

"We should try that, Thoric," Breena said, her eyes shining with enthusiasm.

Thoric snorted almost exactly like Smog always did. Frall and I laughed.

I guess we would have continued in that vein, but a loud buzz interrupted us. Thoric bolted up. "Ah," he said. "We have company."

"More *Old Ones*?" my uncle asked.

Frall said, "No one else can come here. It has to be."

"Well, *we* did, didn't we?" Smog said, laughing, as usual, at his own joke.

It was a young couple, Nelona and Goor, who'd come visiting. Their eyes scanned all of us, but then fastened on me.

"You are Clista," the woman said. "I recognize you."

"From where?" laughed my uncle, but no one answered him. I wasn't sure either. The only time they could have met me was in Herva.

I bowed my head in greeting. "I am pleased to meet you."

The woman walked closer. "May I touch your arm?" she asked. I looked to Frall. He was smiling. He winked at me, so I nodded. Nelona came closer, and taking my arm in hers, she steered me a bit away from the others.

"How strange," Franca said, and I'm sure the others agreed. I didn't hear what else they said because Nelona and I continued walking over to a nearby bench, where we sat down.

"It was a big step you took today. We understand how difficult it was for you."

"Everything is difficult and different," I told her. "I really don't understand any of what's happening. Why have you *Old Ones* called me here? Why does Thoric keep saying you want to train me?"

Nelona laughed softly. "You and I are relatives, my dear. I believe that is what Rwawans would call us. I have been sent to offer my help."

"I thought that was Frall's department," I said, slightly uneasy with her suggestion.

She laughed. "He has tried. I think you need a female's touch. You feel different goals inside him, perhaps. We would understand each other better."

"I'd like your help," I said, and smiled. Nelona was very pretty with shoulder-length black hair. Her bangs curled under slightly, giving her a fifties look. Her vivid green eyes set in the perfect almond shape of models and cheerleaders looked both intelligent . . . and kind.

She was waiting for my response. I nodded with an even bigger smile. "Are you able to answer lots of questions?" I asked her eagerly.

Once more, she laughed. "Ask them, Clista. That is why I have come to you in a Terran's body. Does it please you?"

I nodded. "I feel a lot more comfortable when you look like this and when I'm not surrounded by silver rain."

"Why have you come?" Frall whispered angrily, having slipped away from the others and crept up behind us.

"Do not be angry, queen's mate. You look at me as if I offered a challenge to your possession," Nelona told him with a bit of worry in her eyes. "I do not. I only came to help you and to soften the transition for Clista."

I didn't like her choice of words. I stiffened from the insult in them. No one owned me.

But Nelona hushed me before I could speak. "Easy, both of you," she said. "We are aware of the bond between you. We want it. And, Clista, I did not mean the words to be taken as you understood it. No one owns you. The bond simply means you are mutually tied together."

Frall laughed bitterly. "Yeah, but you guys forgot to include Clista in that. Why couldn't you give her instructive dreams as well? Maybe then, she wouldn't resist me at every turn."

Nelona laughed, throwing her head back as Thoric always did. "Ah, you two. The fires of Miklshew are in your eyes when you look at each other and in your words, no matter how much you dispute it.

"Did we promise you it would be easy, Frall? You woo a queen. That is never simple," she added.

"And what are you three discussing so earnestly?" my uncle asked, barging in most inconveniently.

Nelona stood. "I'm sorry. I do not remember your name."

"I am Clista's guardian and uncle. I am Cristoff," he said, looking sternly into her face. "I guess you didn't hear me when I asked where you'd met my niece before."

"I heard you, Cristoff, but in the confusion, I had no time to respond. I have been with her for many months. It is I who have spoken into her mind."

"You did?" I said, turning to stare at her. "You mean, you're the one who put the necklace on my neck?"

She smiled. "No one placed the *Vremorel* around your neck, my dear. It chooses on its own. It has always done so, and it will continue to choose its path in the future. It sees inside the soul. It measures."

"The *Vremorel*? That's your name for the necklace?" I asked, fingering it as we spoke.

"Yes. The *Vremorel* always picks our queen. For too long, we have been without one. You cannot imagine how delighted we are that it has chosen again."

"That's nonsense," my uncle said as he stood, towering above us, his arms crossed, his foot tapping.

"My uncle is a policeman," I explained to Nelona as if that explained everything.

She didn't seem upset that he'd interrupted. She simply examined him as if he were an outlandish item she'd just noticed for the first time. "How strange that you would say that, Cristoff. I believe you've seen many of the things your niece can do. While she was saving your life from the flying ships of the Cronoks Lab, did you think her actions were nonsense then?"

"That was you? You helped us? Thank you so much," I said.

Nelona threw her head back and laughed. "Don't be ludicrous, my queen. I did not need to help you. You saved yourself."

"Me?" I glanced at Frall and then at my uncle. "I couldn't do that. It was you or the other *Old Ones* who disabled the ships. I called for help, and you guys answered."

Again, Nelona laughed, throwing her head back and slipping into the pleasure of the moment. "My queen, they told me you were humble, but I can barely believe the extent of your modesty. No one needs to protect you, High One. All that you have done has been channeled through you. You have an abundance of power inside you."

"You mean because I wear the necklace? It gives me power?"

"No. The *Vremorel* connects you to yourself. The power was already there."

"That is quite enough," my uncle burst in angrily. "Clista doesn't need to hear more of this rubbish. *Vremorel*, connecting to herself — queen of the *Old Ones*? What are you trying to do, turn her head? She is a perfectly normal kid, and I want her to stay that way."

Nelona called over to Thoric and Goor. "Are you aware that the uncle still thinks Clista is a normal child? He says he wants to keep her that way."

The *Old Ones* laughed. My uncle's face grew redder than Smog with a sunburn. "Clista, come on. It's time to leave," he ordered me, touching my arm as if he needed to get my attention.

I didn't want to go. I turned in the direction that my uncle wished me to go, but I looked back at Frall and Nelona. "I'm sorry," I said. "I have to leave now."

"Nonsense," Nelona said, waving her hand as if she were opening one of the Flagan Pren doorways. As she did, my uncle froze in the middle of a gesture, the one indicating that I should come with him.

"I see what you mean, Thoric," Nelona said. "The uncle is indeed a problem. What shall we do about him?"

I turned to glare. "You will do nothing," I said. "I will not have any of my new family touched."

Nelona, instead of being angry, broke out into another one of her expansive laughs. "Ah, she is a delight, Thoric. The fire of Miklshew springs forth from her eyes. I can see the naked power in her, raw and untapped. How delicious. Yes, she is almost ripe. We must unite them soon. She must pass through Mistah."

"She is not ready," both Thoric and Frall said at the same time.

I had no idea what they were talking about, but I was pretty sure I didn't like it. I certainly objected to their talking *about* me. I frowned and shook my head.

Goor rose from his chair next to Smog. "Clista, look at me," he commanded as he walked closer, his eyes fastened on mine. "Little queen," he said, "let me see your warmth."

I didn't know what he was up to, but I sensed it was something as strange as Herva. I backed away, refusing to meet his eyes.

"Leave her alone," Frall hissed at them as he strode toward me. "Thoric and I told you. Clista is not ready."

"And neither are you, Frall," Goor boomed out. "Do you dare challenge me? Do you even have the knowledge to know what I might do to you?"

Goor's eyes were fiercely piercing. His body seemed to tower over everyone, growing taller and larger in strength as we watched.

"I know," Frall said. "But I won't let you touch her. She is not ready."

It was too late. Despite my attempt to look away, my eyes betrayed me. One glance, and I couldn't free them from Goor's regard. "Frall?" I cried out.

"Touch the necklace, Clista," Frall called out, then moved closer to take my hand. "Hold on to it. Hold the *Vremorel* and force your eyes from his."

But Goor's eyes were flaring. I was entranced and yet horrified. I wanted to look inside his soul. I wanted to run. I could do neither. I was entrapped, my mouth ajar, my body inactive. I knew I couldn't have resisted Goor for even a moment if Frall hadn't stood beside me. But even with him there, I could feel the drain on my strength. Goor's eyes were weakening me.

The necklace was burning the hand that touched it. I felt its flame, but still, I didn't let go. *I'm trying, Frall,* I told him, using our mental contact since my mouth wouldn't open to speak, *but he's too strong for me.*

"He's too strong for you right now, Clista, but he's not too strong for *both* of us," Frall said, and he reached out and touched the *Vremorel*, his hand covering mine.

A ray of light shot forth from the necklace at the point where our clasped hands touched. I gasped. I would have let go then if I could have.

Frall? I cried into his mind. The light was a focused beam that hit and bounced off the far wall, then launched itself at Goor. *What is happening*? I thought at Frall. *What are you doing?*

Concentrate, he told me. *Concentrate and fight.*

"Do it again," shouted Nelona. "Once more!"

The second time, the beam hit the dirt inches from Goor's feet. The man laughed and threw his head back exactly as I'd seen the other *Old Ones* do.

"Good one, children, but not nearly close enough. Feel my eyes, my queen. I am trapping you, channeling you into my power. Do you like the feel of that?" he asked.

I tried to answer, but my tongue was limp, my whole body frozen.

Think of the beam, Clista. Center on it. Then think of Goor. Will you let him take you from me? Frall demanded.

What? I don't understand. Why is he doing this?

CONCENTRATE, Frall screamed into my mind.

Once more, I felt the flame inside me gathering. I formed it and modeled it as if it were a ball of clay, still not understanding but feeling Frall's urging, his directions in my mind.

All right. It's ready. Now send it at Goor. NOW, Frall ordered.

If I'd thought about it, I wouldn't have been able to launch that burst of flame, but Frall's order pierced through my reluctance, and I released whatever that beam of light was. It shot like an archer's arrow, heading directly at Goor. This time, my aim was not short. It knocked into his body, spun him around in a circle, and lifted him several feet up into the air. The *Old One* dropped back to his feet, lost his balance, and tumbled to the ground.

At that moment, I was free. The mind lock was suddenly gone. I stepped back and threw myself into Frall's arms. "You did it," I said. "Thank you. Thank you."

"Me?" Frall whispered. "Don't be ridiculous, Clista. The power of that was all from you. All I can do is guide you. I'm your mate, remember?"

"Good job," Goor said, standing up and dusting himself off. "Most impressive for the first time. I'd say that was enough for tonight, though."

"Congratulations," Nelona said, clapping. "Very nicely handled."

"Are you okay, Clista?" Thoric asked.

"Of course, she's okay. She is our queen. She was meant to do such things," Goor said. "There will be no more innocent modesty from you, my queen," he continued. "You have proven your power. Frall, you'd better help her sit down before she falls. Can't you see she's half-burned out?"

"You'd better not have hurt her," Thoric said as he rushed toward me.

"Hold it. Don't get any closer, Thor. They're still connected," Goor cautioned him.

Frall had been soothing me in the way he knew best. His lips tasted of sweetness. I didn't want to stop kissing him, even when I felt him start to pull away. "No," I cried out. "Hold me, please."

"Of course, I will," he murmured into my ear, but I felt his muscles gathering up to pull away. I tightened my hold around his neck. It didn't stop him. He swung me up in his arms, kissed my forehead, and carried me up toward the throne.

"No! I don't want to go anywhere," I told him so tiredly I could barely argue.

"It's not for that, Clista. You need it for strength. Trust me. Bring Clista something to eat," Frall ordered the others. "NOW."

He placed me down on the throne and scooted in beside me. His arm once more surrounded me, and his lips gently touched mine. I was content.

Meanwhile, Nelona ran to the table, grabbed some grapes, and returned with them. She climbed up the steps of the platform and handed them to Frall. Like a baby bird, I opened my mouth and allowed Frall to feed me the grapes one by one.

It was strange how those grapes were all that mattered at that moment. Hunger chased away every other worry. As Frall placed each grape on my tongue, I was aware of the conversations going on around me. They were discussing what they'd seen. I pulled in their words and filed them for later inspection, but for the moment, my eyes and interest were only centered on Frall and the grapes.

"I think I see," I told him.

"Don't talk yet," he warned me. "Darn Goor for pushing you."

"I'm okay, Frall. Eat some, please. You must, also."

He popped another grape into my mouth and tossed one into his. Then he went down on his knees. "How much do you understand, Clista? How much?"

"You're not proposing again, are you?" I giggled stupidly.

He smiled. "Not unless you want me to. Do you?"

"No. I want us to stay as we are, maybe another year. Can we do that?"

Frall's eyes studied mine. "They won't like it. They will keep pushing, but I will fight them, if that is your wish."

"You fought them tonight for me. I don't know what that was all about, but it was only because of you that I withstood Goor's eyes."

"Yes," Frall said, confirming it.

"Do you mind so awful much, Frall? Will you still be patient with me?"

He laughed, not like the *Old Ones*, but with a gentle laugh, the laugh I'd come to love. Then he touched my lips with his. "Yes, I'll mind, but I'll be patient, as patient as you need me to be, Clista, because I love you. You do understand that, right?"

His eyes entreated. His face looked so earnest, it was almost like Goor's in that moment when I couldn't pull away. Only I recognized the difference. I wanted to gaze into Frall's eyes. I loved doing so.

I nodded. "Thank you. I think you're wonderful, and I love you, too."

"You understand that when we're together, it multiplies your power?"

"I remember that, but I don't understand why or how. And I really don't understand what Goor was trying to do. Why did he do that?"

"He was testing you, Clista. He wanted to see if you really had the power inside you."

"But I could have hurt him."

Frall laughed. "You sound like yourself again. I don't think you'd ever hurt anyone, Clista. You don't have that in you. I bet Goor knows that, too."

"Are you all right?" Thoric asked. "Do you need something more?"

"We need another meal, I'm afraid," Frall said.

"Of course. I'll bring some sandwiches."

Nelona and Goor crept closer, their eyes on us. "That was quite exciting," Goor said. "You pleasantly surprised me, my queen."

I glared at him. I didn't want to tell them that I didn't understand again. That cry for help was always ignored, but I certainly wasn't feeling friendly. He'd frightened me. Why had he done it? Not knowing exactly what to say to him, I remained silent. It was just as well because Frall had plenty to say.

"You could have harmed her," he yelled down at them.

"Harmed her? Of course, I would not do anything to hurt her. She is our queen. I was extremely careful with her."

"She's all shaken up. You think that was careful?"

Goor stared at me a moment, then blinked and looked away, his handsome face wreathed in ego and an almost overwhelming sense of egoistic grandeur. Like Nelona, Goor's hair was ink black, his eyes the exact same hue of green as hers, the pores of his skin equally radiant with vitality and . . . something else that I couldn't determine. Good health, or was it power, the power of the *Old Ones*?

"Really, she is taking in everything adequately," Goor said.

The two of them were studying me like I was an amoeba under a microscope. I stared down at my lap. I wasn't about to meet Goor's eyes — not ever again.

"Tell me, Frall, how much was she holding back? Was that all of it, or was there more?" Goor asked.

That was the final straw. My temper flared. My eyes did, too. "Enough. Stop it. I am not a *thing* you can talk about. I'm a person, and I'm tired of all this. Leave me alone, Goor. You, too, Nelona. I don't want to be your queen. Now stop talking about it."

Thoric was bringing over a platter of sandwiches. He almost dropped them. "Clista, you don't mean that," he scolded.

He turned to glare at Goor. "See, I told you she was not ready. Now look what you've done."

"She is a child," Goor said, untouched by Thoric's criticism. "What else can you expect? But she needs to outgrow that and start taking up her responsibilities. We cannot keep catering to her sensibilities all the time,"

"Hush. You're making it worse," Nelona said. "Look at her. Her eyes are blazing. She's such a beautiful queen, and potentially . . . well, she is obviously dangerous when provoked."

I bolted out of my seat, tearing myself away from Frall's arms. "Bring my uncle back. Unfreeze them now and leave."

Frall stood behind me, draping me with his arms. "It isn't wise to anger them, Clista," he whispered. "Goor is one of the most powerful of the princes. Besides, this is Thoric's house."

"Goor's a prince?" I said, stumbling over the word. There was nothing about him that was in the least bit princely except, I supposed, his cocky attitude.

I hear your thoughts, my queen," Goor said as he watched me with his eerie green-hued eyes.

"Get out of my mind," I ordered with a new firmness and certitude that I could demand such things.

"Clista, Clista," Thoric warned, handing me the platter of sandwiches. "Calm down. Do you need to lie down for a bit? Are you truly okay?" He glanced at Frall like he was the better judge of my situation.

"Clista. Tell Goor you're sorry and do it quickly," Frall whispered in my ear.

"I will not," I said, flaring up at him. "I don't like the man or whatever he is. Ghost being, spirit, *Old One*. I don't care anymore about appeasing these aliens. I've had too much of their craziness, and I'm not going to take it anymore."

"Good. Then what will you do about it, my queen?" Goor dared me, taking a step closer, once more threatening me.

I cringed, scooting back into the throne, although my movement was no more than an inch. Besides, I knew that there really wasn't any place to retreat. Not from an *Old One*.

"I shall banish you," I said. "When I close my eyes, you will disappear, and when I open them, you'll have been transported elsewhere," I stated firmly, hoping it would work but not quite believing.

Frall was always telling me that I should believe in my power. Well, maybe I would. Maybe such things really were possible. I closed my eyes and imagined that. I visualized Goor disappearing, fading out into nothingness. I saw him returning to his world of Herva and swimming about in the sea of floating silver drops.

Then, I pictured myself opening my eyes and not seeing him in Thoric's house anymore because he was suddenly gone. It was a good image, and I reinforced it by going over it several times. Then I gave it to the count of four and touched my necklace. One, two, three, four."

The necklace didn't warm to my touch. Yet, I still half believed it would work. The room was silent. I peeked with one eye.

Goor's face had broken out into silent laughter. He was convulsing with it, but he wasn't disappearing.

When he saw my eyes open, expecting him to have disappeared, he let out a burst of full-voiced hilarity and dropped down in a chair where he could have his hysterics.

"She is delightful. Such a mixture of innocence, ignorance, and unrealized power. I absolutely adore her, Thoric. You must invite the others to see her. The princes, at least. They must. They must observe her in her present state of sweetness and purity. They will relish it, recalling it for centuries to come."

"Oh, stop it!" I said, stupidly stamping my foot. I turned to face Frall. "I don't suppose you can get rid of him, can you?"

Frall shook his head. "I'm still learning, too, Clista. Thoric told you that I know all about the power of the *Old Ones*, but I really don't. I see the broad picture, and I know that Goor is an influential player, but that's about it."

I sighed heavily. "Then I guess we just have to ignore him, right?"

Frall shook his head. "I've been told that the more we go to Herva, the more it will all make sense."

"How could that be? All I discovered in Herva is their real form…"

"Their spirit form, Clista. All forms are real to them," Thoric corrected.

Thoric still remained in front of us, holding the platter of sandwiches. I took it from him and handed Frall a sandwich. "Thank you, Thoric," I said. "I don't suppose *you* could make Goor disappear, could you?"

He shot a glance back at Goor and then winked at me. "Ah, Clista, you just do not yet understand how to handle him. Maybe it would be a good idea if we all sat down and talked about it."

I shrugged but followed him back down to the table, carrying the platter of sandwiches. Frall took the seat on my right. Thoric almost sat beside me on the other side, but Nelona slid into the chair before he could do so. He ended up sitting across from us, beside Goor and next to Smog, who was still frozen, the latter's mouth pursed like he was in the process of asking Breena a question.

I had decided not to speak. I would let Frall carry the discussion. I was too upset over Goor's laughter and his treatment of me. I bit into a sandwich and tried not to look at the *Old One* prince.

"All right, children, ask away," Goor said, taking the lead once more.

"What was the purpose of what you did to Clista?" Frall asked.

I gave him a smile for that question as I continued to chew.

"As I said," Goor said, tossing a grape into his mouth. "I was merely testing her. We wouldn't want a weak queen, would we?"

"But Nelona said that the necklace chooses the queen. Why would you need to test Clista?" Frall continued.

Boy, Frall must have been reading my mind. I took another bite of my mashed soy-ham sandwich and poured myself and Frall a glass of lemonade.

"The *Vremorel* chooses, but it does not tell us the qualifications or characteristics of the queen it has chosen. How can we know what her structure is if we do not probe a bit?"

"Why is it necessary for a queen to have power? Do you have enemies she is supposed to fight?"

"Frall," I interrupted, swallowing the bite I was chewing. "There's no way I'm going to fight anyone."

"Sh!" he told me. "I thought you wanted me to ask the questions."

Had I said that? I tried to think back, but it didn't matter. Frall could have just picked it out of my mind.

"One always has enemies, my boy, but none she needs to fight. Our enemies are monotony and longevity."

That set me off thinking about the fact that neither couple seemed to have children. How long did *Old Ones* live?

"Where are your children?" I asked.

"Children? Why, you and Frall are our children, of course," Nelona said, looking slightly puzzled.

Thoric poured himself a glass of something he hadn't offered us. He passed the decanter to Goor, who poured for Nelona and then capped it.

"I think," said Thoric, "that it's time for them to have another bite of information." He glanced at Goor and then at Nelona.

Goor shrugged. Nelona asked, "Are you sure, Thoric?"

Beside me, Frall put down the rest of his sandwich and frowned at them. I swallowed and prepared for more bad news.

Thoric's eyes stared into mine. "You asked, Clista. It's possible that this will be hard for you. And for Frall."

"Stop avoiding it and tell us," Frall said, far braver than I was because, looking into Thoric's eyes, I wasn't sure I wanted to know.

Thoric's eyes studied Frall. "You see, you are my grandson, Frall."

"No way." Frall lurched up from his chair. "I knew my grandparents. Both of them. They were scientists with the Druster Mission."

"You knew the humans in that role. I was your father's father, Frall, not that man you thought of as your grandfather. Of course, he didn't know the difference. In fact, your grandmother didn't know either. We often do not tell our host human."

Frall paled. "You mean that you raped my grandmother?"

Thoric laughed and shook his head. "Not at all. When we choose a human, we only borrow the form of their husband or wife for a while. Then we deliver what we need to give and journey on. It is our way. We can no longer bear children among us. We are too much spirit and not enough flesh."

"When you told me before that I was part *Old One*, I thought you meant that my grandmother was an *Old One*. You didn't explain it well," Frall complained.

"Your grandmother was half *Old One*. We did not lie to you. You are the third generation. Many generations are necessary to strengthen the line. You see, the power does not show up until the process has been repeated three times."

"I am your grandfather as well, Frall," Goor said. "I am your mother's father."

"No way. You can't be. You're too young," Frall stuttered as if he were unable to process so much at a time. "How can you both be my grandfathers?"

The fact that Frall was related to Thoric made a great deal of sense to me, but something told me that I, also, would have my reckoning. I knew with certainty that I wasn't ready for that. I wasn't sure I even wanted to know about my family.

"Your mother was my daughter," Goor said. "Your father was Thoric's child, and both of your grandmothers were the children of *Old Ones*."

"We have carefully planned your arrival, you see," Thoric told Frall. "We have watched you and your parents and your grandparents for all these many, many years. Your heritage is more attuned with us than with those of Rwawan.

"Had Cronoks Labs not unearthed the *Tworst of Twent*, we would have allowed you to sample a bit of adulthood before revealing your history and claiming you for Flagan Pren."

"What about Clista. Is hers the same story?" Frall demanded, as if he'd already accepted his own tale.

Thoric's eyes darkened. "Read her, Frall. Is she ready?"

I lowered my eyes, not wanting to be the center of their conversation. Besides, I had no intention of participating. I wasn't about to accept anything they said. It was all lies. They were trying to confuse us.

"Clista," Frall said softly, pulling my eyes to him despite my resolution. "They aren't lying. It all makes sense now. I feel it. I know it. Believe them. It is the truth."

I shook my head. "You want to believe that they tricked your grandparents? That's a good thing? It's deceitful. It's wrong. And it's all lies, Frall. It never happened."

Nelona reached out a hand to me. "I am your mother's mother, Clista. Do you doubt that?"

I stared. It's said that the eyes are the windows of the soul. Does that count with *Old Ones*? Do they have a soul when they spawn children through trickery and leave them for others to raise?

I removed my hand from hers. "My grandmother was Betsa. She died of old age. And she wasn't an *Old One*. Neither was my mother. She so badly wanted to be, though. Your story would have pleased her, but nothing lit up for her, none of the artifacts. That proved that she had not one drop of *Old One's* blood. She cried because of that. And now you say she was half? No, I don't accept your lies. Not any of them."

"Pour Clista a drink. Just a small one, but the kid needs something. She's going into shock again. I feared it was too early," Nelona said.

"I don't need a drink, and I'm not going into shock. I just don't choose to believe your lies. Leave my parents and my grandparents out of this conversation. You'll never convince me. Not ever," I told them.

"Why do the others sleep while you sit here talking, then?" Goor asked, leaning forward, his smile a sneer.

"Because you're cruel, and it amuses you to be," I said, replying before I'd thought to soften the words. But I was too angry with the current conversation and with what Goor had done earlier to worry about politeness."

"You will not punish her," Frall said, bolting out of his chair to stand in front of me.

"Sit down," Goor said, shaking his head and chuckling. "If I meant to punish the queen, your standing in front of her wouldn't stop me."

I reared up my head, wishing I was a mythological dragon and could puff fire into Goor's face. "Frall, quit protecting me all the time," I said. "I can defend myself." I shot that out before I thought really hard about it. Then I flushed and looked down. "That's a lie, isn't it? I'm sorry, Frall."

He turned to face me. His hands reached out to cup my face. *Be careful. I have already explained that Goor is one of the high ones. He can hurt you, Clista.*

I suppose I would have argued or told him I didn't care, but Frall abruptly let go of my face and sat back down. Had he been ordered to do so? I studied the faces all around me. If the *Old Ones* all declared that Rwawan was a planet or that its sky was bluish gray, I think that despite my knowledge that those were truthful statements, I might have suddenly doubted the veracity. I was that mistrustful of them at this point.

Nelona took my hand again. "The other *Old Ones* of your family have not yet come, child. They could have. They would have if they'd realized we would begin our tellings today. But they wanted to give you more time to assimilate.

"I think they were right about that. Tomorrow, we will discuss your heritage, my dear. Your family will be here then, and you will meet them. Then it will be okay. You'll understand and accept.

"But trust me for now, Clista. I won't lie. I carried your mother in my womb for nine months, just as a human would. I am truly your grandmother."

I pushed my chair away from the table. "Please, I have to leave. Frall can stay and talk to you. He obviously accepts this garbage, but I need to go to my room. I can't take anymore. I need to think."

Frall stood up.

"No, stay here," I told him. "I don't need you to go with me."

"I won't let you go by yourself," he said, throwing his arm over my shoulder as if I were a child who needed tending.

I wanted to throw it off and shrug away his over protectiveness, but I just stood there, mute, not taking another step.

"I will escort Clista," Nelona said, taking my hand again, relentlessly persistent.

I pulled away from her and from Frall. "No. I want my uncle and Smog and Franca. That's all. Just them."

Thoric nodded. "You know who you remind me of right this moment? Stav. She is the perfect image of Stav."

"You are right," Goor said with a laugh. "The facial expressions, the stubbornness, and the determination to stand alone . . ." Goor howled his laughter up to the stars while I stood there, burning red, waiting to hear if they would allow me to leave and if they would wake my uncle and the others.

Nelona sighed and returned to her chair. I looked down at her, wondering if I should apologize for my accusations and my rejections, but I couldn't. Besides, she didn't look wounded. In fact, she was laughing, too.

"Stav will be here tomorrow. He will be delighted with her. I know he will see the resemblance — that old battle warrior. He really scored true this time," Gore said.

I sighed loudly enough for them all to hear my impatience and irritation.

"Please, let her go to her room," Frall said. "She'll step forward if you give her the time to think it over. But you can't keep poking her or thrusting new things at her."

"Of course, we will continue to push her. We are not content watching you sit back, unable to play your role. She has to be pressed," Goor told them.

"Hold it," Thoric said, his hand raised as if he were directing traffic in the old vids of Earth.

Everyone stopped and looked at him. "I agree with Frall. Clista, I will free them, but remember not to talk about this. It would only cause trouble. Even Franca, who is sympathetic to our power . . ."

"I bet there's an ancestor somewhere in her line," Nelona interrupted.

"There are no records of such," Thoric said, forgetting to continue with his thought.

"But she is connected, as you said. She feels the current. She must have the blood," Nelona continued.

"Nelona, my dear, allow Thoric to continue, if you will," Goor ordered her with a special softness in his voice that had not once been directed at Frall or me.

She sighed and picked up her goblet to sip daintily. "See how it is with mates, Clista? They are delightful to have most of the time, but sometimes, they do get rather dictatorial." She laughed and smiled so sweetly in Goor's direction that his eyes tempered, and he smiled back at her.

"You were saying, Thoric?" Nelona urged, giving in to her husband's request.

Thoric shook his head. "My wondrous lady, your charm enchants. I did forget what I was saying."

He continued at the exact point where he'd left off. "Clista, as I started to tell you, you must not discuss this even with Franca and especially not with that uncle of yours. Neither he nor Smog has an iota of power. They would not believe, and they would indubitably double the watch they keep on you, restricting you further from enlightenment."

He glanced at Goor and Nelona and waited for a nod. When he got it, he waved his hands, and Smog, Breena, Franca, and my uncle all came alive. My poor uncle, standing over by the bench where we'd been sitting previously, was left with his finger in the air in the middle of a sentence in which no one was at the bench.

"What happened?" he asked, looking mystified that we'd all moved in what must have appeared to him, a magical method of teleportation.

I dropped my eyes, trying to think up a suitable excuse without telling more than Thoric had warned me to say.

Nelona smoothly inserted, "Why, nothing. We were just sitting here chatting about things. Come join us, Cristoff. Here, have some wine.

"I offered a bit to your niece since we have no rules about that here, but she declined. She is such a good girl. Smart and eager to learn. You must be very proud of her."

My uncle looked like one of those freshwater trout that Smog had caught for our dinner, its gills flapping open and closed, trying to suck in air from water that was no longer there. But then my uncle's mouth froze into a rounded open, and still, no words came out. At last he sat down, poured himself a glass of wine, breathed in the smell of it, and tasted.

"Very nice," he said, "and yes, I'm quite proud of Clista because, so far, she hasn't let any of this ridiculous queen stuff go to her head."

I was still standing, ready to go, but it didn't look like the *Old Ones* were going to release me that easily.

"Franca, I'm going up to my room," I said. "Are you ready yet, or do you want to stay a while?"

"Really, Clista. That's rather rude. We've hardly had time to visit with Thoric's guests," my uncle chided me. "You should stay and be polite. You might learn something."

"I know. I'm sorry, but I have a headache. Perhaps I will see them tomorrow."

I apologized to everyone, even though half of the group knew exactly what my problem was. I pretended, as Frall had taught me, that they didn't.

I pushed in my chair and set off, walking toward the elevator.

"Wait, I will go with you," Nelona said. "I would hate for you to get lost. Thoric's house is a maze of floors and rooms."

"I have probably been here longer than you. I'm sure I can deal with it," I told her. That received a sharp admonishment from my uncle and a frown from Smog.

"I didn't know that a headache interfered with courtesy," Smog chastised.

"You don't understand," I snapped, but then I saw the look on Thoric's face. I bit my tongue, and needless to say, Nelona walked me to the elevator, got in, and punched the coordinates.

"I could have done that," I said, irritated by her persistence.

"Of course, you could, but how often have I had the opportunity to do it for my little granddaughter?"

I hissed and looked out the window as we lifted through the floors. When the elevator stopped, I stepped out, trying to escape her, but she didn't accept it.

"Show me your room," Nelona ordered.

"I don't want you to go with me," I told her rudely. Brazenly, I stared into her eyes so she'd see my feelings about the subject.

"I know, dear, but I am going with you anyway," she said gently.

"I need time to be by myself," I tried again.

She laughed. "You and I are going to have some chatting time, headache or not," she said, laying her desire out in the open.

I halted. I'd tried hints. She wasn't getting them.

"No. You're not welcome to come into my room," I said at last. "Go back with the others. I was trying to be nice, but the truth is that I've had enough of your lies and your tricks."

Chapter Seven

"Dear child, then you give me no choice." So saying, the *Old One* raised her hand and brought it over my head. I couldn't help looking up. I was just in time to see her click her fingers. Then, the world began to spin. Dizzy, I shut my eyes.

When I opened them, I found myself outside and all alone. The dry dust of the prairie-like sand was everywhere, and the wind was blowing fiercely. The leaves of a nearby tree rustled and were tossing about like angry fingers, scolding me. I turned around to see the rest of the landscape and to search for Thoric's house, but it had disappeared. There was nothing in sight except the tree and a wide expanse of yellow-brown dirt.

"Why have you done this?" I called out, but there was no answer. The wind continued to howl and blow, and the dust attacked me with a fresh barrage of tiny stone pellets. I covered my eyes. Then I dropped to the ground, sitting in the old crisscross applesauce posture. There was no place to run to and no shelter from the wind's buffeting. The tree was at my back, but its leaves offered little protection from the silt-like powder that was blowing all about. Soon, its dust caked my nose and covered my face.

Why had Nelona done this to me? Was I being punished? She was Goor's wife. Had she been insulted when I'd said what I did to him, calling him evil? Or when I'd been nasty to her, refusing to talk. Yet, she claimed she was my grandmother. What kind of grandmother would do this to her granddaughter? And how had she done it? I'd been inside the building a minute ago, a building that had vanished.

It was not cold outside, although the wind blew with gale-like proportions. Nor was it a hot wind, but rather, neutral, only irritating. I tore two strips from the bottom of my dress, using a twig from the tree that had a nice, sharp point. One piece of cloth I used to tie around my forehead, sliding it down, so it covered my eyes. The other, I placed around my mouth and nose so I could breathe without inhaling the dust.

Tying both pieces at the back of my head kept my hair out of my face as well, which was a considerable relief since it had danced about and thrown itself at my eyes with stinging force.

I sat under the tree for several minutes, hoping to hear someone calling my name, but there was only the wind screaming its forlorn call as it battled with the dust and flung itself at both the tree and me. The currents of wind made ripples across the small hill-like dunes.

"Snake dances," I could almost hear Frall saying. That's what he would have called it if he'd been with me, but he wasn't. I sat and thought about him and about everything that had been happening. I tried to shut out the loneliness of the place — and my fear.

Perhaps I sat there for an hour. Maybe it was a much shorter period of time, but at last I realized that no one was coming. I stood up and fought the wind. I had no destination that made any more sense than another, but I set off, needing to move. My feet seemingly made little progress as the sand slid out from under them at each step, but I continued, climbing the first mound and then another and another.

I came across no more trees, and when I looked back, I couldn't see the place from which I'd started. Then my journey seemed even less meaningful, but I kept walking, one foot in front of the other, sliding back, falling on occasion, but constantly striding forward.

In the distance, through the dust-riddled air, I saw another tree, so I made my way toward it. It didn't look that far, but it took me forever to bring myself closer, one step at a time, over and over. At last, I came to it and collapsed on the ground, hugging the tree as if it were a good friend. I rested a bit and closed my eyes. They were burning from the sand and dust, and I'm sure they were red from the dryness of the wind.

The shrieks from the air currents didn't lessen, nor did they stop buffeting me with sand and dirt. Eventually, I opened my eyes again to look around me at my new location. It was then I discovered the shreds of material, still caught on the twig I'd used to tear my dress. After all that walking, I had only returned to my original spot.

I cried then. I suppose that was good because it cleaned the dust from my eyes, but it offered no relief. If anything, my eyes burned twice as much, and the disappointment I felt at having learned that I'd just walked hours in circles was a further instrument of misery. I threw myself down in the dirt and mingled my screams with the winds.

Suddenly, I heard the snap of a finger, and the wind died instantly. I opened my eyes. I was standing in the hallway in front of Nelona.

I stepped away from her and turned to run, but my feet wouldn't move. The soles of my shoes were dried cement. I could not lift them.

"Stop it!" I cried. "Why are you tormenting me?"

"Because you are acting like a child, Clista. If you choose that role, I will provide lessons."

"You call those lessons? You left me alone in a deserted area, and the wind battered my skin. I was almost blinded by it. My eyes still hurt."

"Your eyes hurt because you were crying," she huffed.

Tears were streaming down my cheeks again. My body felt like lead, and I was suddenly so thirsty I would have given anything for a cold drink. "Please, let me go. I'm sorry I said what I said, but please, let me go."

"I do not think you have learned the lesson yet. I think you need another."

She raised her hand over my head, and once more, I found myself spinning dizzily. The motion of my gyration caused my stomach to complain. When I stilled, I opened my eyes, certain I would vomit. But I didn't. I was too shocked by the sight. I gasped. I was surrounded by water. My arms and legs began to paddle.

I understood that I was being punished again, and that it was probably all just an illusion, but that didn't change anything. When I didn't keep kicking, I started to sink, and the water was as real as the sand had been.

I dog paddled around in a circle, searching the horizon for land. There was none in sight.

"What do you want me to learn?" I called out, but there was no answer. "I'm sorry," I screamed, but the only sound was the splash of my arms and the tiny waves slapping at my face as my feet battled the water.

I dog paddled on and on, wondering when Nelona would call me back, but at least an hour passed and still I remained. It was exhausting to fight the water. At last, I grew so tired I just let go and sank, but the water down below was icy, as if the sun, wherever it was, had heated only the topmost layers. I shot back up and found the energy to tread for a little while longer.

My arms ached. My legs had developed cramps, which I kept massaging and trying to soothe. At last, I sighed heavily and lay back in the water, hoping that I could float for a while.

Spreading out across the surface, my arms stretched wide and my legs v shaped, I rode the waves and listened to the sound of the water. It soothed me. I floated about like a cork on a small brook, the water lapping at my face, kissing me — just as Frall had when he'd lain at my side during the nights of our trek through the forest.

"Frall," I called. "I miss you. I really do love you, you know."

"Now, that was the first sensible thing I've heard from you since we came up in the elevator," Nelona said. I realized then that I was, once again, standing in front of her, back inside the hall.

I let out an enormous sigh and stared down at the ground.

"No sharp words for me?" Nelona asked.

I lifted my eyes to stare at her, but I didn't speak. I was too tired to argue, and besides, I was afraid she'd send me somewhere else.

"Good. I think you're ready for that chat now. Where is your room?"

So, Nelona had her way after all. I sat on the bed, and she made herself comfortable in one of the chairs. Then she questioned and lectured. I reclined with my back against the wall and attempted to listen, but pretty soon, I fell asleep. She must have left then, placing something on the pillow beside me as I snored away.

When I woke, that was the first thing I saw — that and Frall asking me what it was. I sat up, hoping I hadn't been drooling.

"How did you get in here?" I asked.

He laughed. "Your room is locked to others, but not to me. Didn't you know that?

I laughed. That sounded so typical of the way the *Old Ones* thought. I wasn't surprised.

I sat up to inspect the thing that lay on my spare pillow. "I have no idea," I told Frall, turning it around and around. "Do you think it's an artifact?"

"Let me see."

His face was full of curiosity. His eyes had that glint they got when he found a puzzle he was dying to solve. He licked his lips as if the thing Nelona had left was a box of chocolates, although I don't think Frall was the kind who went for sweets. He was more of a burger and fries man.

I handed the strange metal, hand-shaped object to him and then moseyed into the necessary room to use the facilities. When I came out, Frall had already figured out the artifact's purpose. "Look," he said. "It has pictures in it. Watch the wall."

I shouted out the moment Frall flashed the first image. "That's my grandmother!"

Frall kept going. The next one was of my mother as a child. I recognized it. Then, another was farther back in time. I'd seen a picture similar to it before. I was pretty sure it was my great-grandmother.

"I can't believe it," I said, dropping down onto the bed. "Why would Nelona leave me a photo album? I mean, it's strange. She couldn't stick around to meet her daughter and raise her child, but she could take pictures of her? And why would she give this to me? Am I supposed to just accept everything now? *Oh, how wonderful to meet you, Grandmother. Thank you for deserting my mother and making all my relatives a lie.*"

"Clista," Frall said, trying to make me stop.

But I wouldn't. I wasn't through at all. I was watching the pictures as they flashed on the wall, yet I was hardly seeing them. I was too angry.

"Ah, Clista. Give it a break," Frall said, dropping into the chair across from my bed. "It wasn't that way at all. You had a grandmother who loved you. You were taken care of. You weren't just deserted."

"What do I care? A photo album isn't proof, anyway. Nelona's still probably lying about the whole thing."

Frall propped his feet up and shook his head. "Why would she lie, Clista? Why would she want to say she was your grandmother if she wasn't?"

I stood up and began pacing. Frall turned off the picture show and watched me, waiting for me to get the anger out of my system.

"I don't know," I told him. "Maybe my disbelief doesn't make any sense. Nothing in Flagan Pren does, actually. It's all bits and pieces that wander about, colliding off each other like balls on a wall."

Frall laughed. "If you used a little bit of open-mindedness, Clista, I think it would make a lot more sense."

"Not really. Listen, I need to tell you what Nelona did to me. You won't believe it, except, I guess, it's no weirder than anything else."

Frall readjusted his feet. He was sitting just like my uncle and Smog had sat, legs hoisted up in the air. I wondered if that was a male thing, or was it just the fact that he'd watched them do it and was copying?

"I didn't want her to come to my room and talk…" I explained.

"Clista, why not? The woman is your grandmother, no matter what you think!"

"Are you going to listen or not?"

"All right. I sure hope you didn't make her mad."

"I did. She sent me outside, Frall. *Twice*."

"Outside? Do you mean downstairs or outside? Like where we came from?"

I nodded. "She made me stay out there all by myself, and the wind was blowing something awful. I had to tear my dress just so I could breathe, and then I walked and walked, but no matter how far I went, I just kept coming back to the same place."

Frall stood up and came over to sit by me. "I don't get it," he said. "How did you get outside, and why did you stay there? Why didn't you just come back in? You know no one would have wanted you out there on your own."

He'd taken my hand in his. I closed my fingers around it, thankful for the comfort. "It wasn't like that, Frall. Nelona sent me someplace — with the click of her fingers. And that place wasn't anywhere near here. There was only a tree and sand dunes. I couldn't even see this building or the ship we came in."

Frall scooted closer, pulling me into his arms. "Darn, it, Clista. I never would have allowed that," he said. "What was she thinking? I suppose it was only an illusion, but you believed it, didn't you?"

"If it was some kind of magical deception, it was so vivid I tasted the sand, and I could barely breathe because of the dust in the air. I tore my dress. See." I looked down, but there was nothing to see. My dress was whole.

I burst into tears then. It was stupid. I never used to cry, but what else can you do when frustration slaps you around like a windblown sheet?

Frall hugged me. "Hush, you're okay now, Clista. I'm here with you, and I'm going to be from now on. No more arguments about it. I don't like the things they've been doing to you."

"Nelona sent me somewhere else, too. It was like the ocean or a gigantic salt lake. I swam and swam, and there was no land anywhere. I was so tired, Frall. I even tried to give into it and sink."

"That's enough, Clista. Never again. I won't let them."

"Clista!" It was my uncle's voice. I broke away from Frall and looked up. Uncle Cristoff was standing in the doorway, and from the way his face looked, he was good and angry. But he had no reason to be. Frall had left the door open, and we weren't doing anything wrong.

"Clista, I told you that I didn't want you and Frall . . ." my uncle started off his rampage.

"Knock it off," Frall snapped at him. "She's upset about something, and if you can't see that, then you have no business coming down on me."

"Hey," Smog said like that was supposed to mean something.

But Franca was there with them, talking quietly, doing her best to calm down the situation. I hoped she could. Uncle Cristoff looked really, really mad.

Smog marched into the room as if he were taking over. He plopped down in the chair where Frall had been sitting. At once, his feet propped up on the table like he was resting in an easy chair.

"You need to be respectful, young man. You don't want to get Clista's uncle uptight, believe me."

"Yeah, well, then he needs to back down a little," Frall growled, not giving in an inch. "Tell him to give us some space. Clista's my girl, and she's going to be my wife. He knows I'm respectful of her, so he has no reason to be on my case all the time."

Smog gave Frall a measuring eye. "Maybe he's just trying to show Clista he loves her. Besides, he's new to it. He doesn't have much practice, remember? Suppose you have a daughter someday. You gonna let someone manhandle her all the time?"

Frall didn't like Smog's terminology. I had to stop him from swinging a fist at the big guy. Meanwhile, Franca and my uncle had concluded their disagreement and were focusing back on me.

"You feel better now?" Franca asked me.

"Yes. My headache's gone. Thank you."

"Nelona came down whistling. You and she must have had a nice chat," my uncle probed.

"Yes," I said, giving Frall a look. He understood.

"Listen, Clista," my uncle said, looking a bit sheepish. "I'm sorry about coming down on you so hard. I was just disappointed. I thought you and I had made a bargain about Frall. Didn't we determine that he wasn't supposed to come into your room when you were alone?"

I looked down at the floor. "I don't think I can keep that deal, Uncle Cristoff. Frall and I sometimes have things we need to talk about. Sometimes, they can't wait. You'll just have to trust us."

My uncle's face reddened. "I see. Maybe Franca and you should share a room for a while, then. Remember that you're still a minor under my supervision."

"Cristoff!" Franca admonished him. "Don't even suggest it. Clista has handled herself remarkably well for someone her age. You should be praising her, not taking away privileges."

My uncle turned his fiancé around so he could look into her eyes. "Franca," he said. "I love you, but Clista is my responsibility, and…"

"What's this? A family debate?" Thoric said as if he'd just been passing by. I doubted that. Did he have our room wired?

"I was just suggesting that Clista and Franca might want to share a room for a while, because . . . "

"Absolutely not," Thoric said. "If you distrust your niece, maybe you should put a bell around her neck. That might help. Or maybe a chastity belt would be better. I've heard those were popular once on the planet of your origin."

My uncle started bubbling over. Maybe that's the wrong analogy for his rage. Maybe it was more like smoke coming out of his ears, but you get the picture. He was a pawing bull, ready to gore someone.

However, Smog didn't take Thoric seriously at all. He burst out laughing, as did Thoric. Frall and I both blushed. It was so unfair to have everyone constantly making our relationship their business.

"You think so?" Thoric said, looking pointedly at me as if he were answering me. Had he read my mind? Thoric suddenly waved the others still and stared at the two of us. "You know, Clista. I believe you are right. Of course, that is the answer. Forest, desert, mountains, or ocean?"

Frall and I exchanged a look. We had no idea what Thoric was talking about. Apparently, Breena did. For some reason, she was still unfrozen. She chose our fate. "Forest, of course, dear. Remember how much fun they had there before?"

"Ah, my love, how perfect," he replied.

With a click of his fingers and the same whirling dizziness as Nelona had treated me to, I was spinning round and round. I closed my eyes and wondered what I'd done this time. Was Thoric punishing me, too?"

When the gyration stopped, and I found myself on the ground, I opened my eyes. Breena's choice had been Thoric's final decision. I was alone in a forest. I backed up against a tree and sat there, hugging my knees, wondering how long I'd be left this time.

There was a stillness in the air. I recognized it, forest tranquility. Even the breeze didn't rustle the leaves in the deepest part of a forest, the part where the trees grew into the sky, and the ground was padded with several inches of redwood dust.

Would the forest cats, the jasmire, hunt me as they had before? Or would they back away this time? I was thinking about their luminous green eyes when the bushes moved.

I searched for a heavy stick. There were none — not close, anyway. So, I stood up, ready to confront the jasmire with only my bare hands and hopefully a bit of magic.

"Clista," a voice called.

"Frall?"

The bushes shook even more. I screamed, thinking once again that it was a jasmire, but it was only Frall. Of course, I ran into his arms. I hugged him as if I hadn't seen him for a hundred years. "They sent you with me this time. But why?" I asked. "Are they punishing you, too?"

Frall laughed, but it wasn't his good laugh. This one sounded bitter and cold. He kissed my cheek and then pushed me away, almost roughly. "Come off it, Clista. You know why."

I thought at first he meant that I knew why they were mad at us. All right, so I'm slow. I stared at him, not understanding, thinking he meant that it was my fault and that I should know why we were being penalized with this isolation. It was only when I saw him blush that I understood. Then, it hit me. "You mean they want us alone — together? So we'll . . . No," I said, backing away from him.

Frall didn't stop me. He turned and looked up in the sky. "Darn you, Grandfather. Stop interfering."

There was no answer beyond the echo of Frall's voice fading away across the forest emptiness.

"I'm sorry," I said. "It's about me again, isn't it? They're still pushing."

Frall nodded. Then he sighed. "It's okay. It's not your fault. Yes, they're still urging us into this relationship, but we don't have to listen. I love you, Clista, but I won't take advantage of it. You know that. You know you can trust me."

His green eyes were staring into mine. More than at any time, I saw the greatness that was Frall.

"I love you," I told him. "Thank you."

He laughed, but it sounded like his usual laugh. He wasn't mad or anything. In fact, he gave me a funny smile and said, "Come on. Let's go get some dinner. I don't know how long they plan to leave us here, but I guess you've already figured out that the *Old Ones* control time. They could keep us here for an entire month and bring us back to the exact moment we left."

"How can we fight them, Frall? They can keep doing this, sending us here and there and putting us into danger and…"

"We can't fight them. We can only relax into it. It doesn't matter what they do or where they send us. We can have a good time. You're with *me*, remember? It's not like when Nelona sent you off by yourself. Don't worry about them. They can't change us. That's the important part. They can't undo what we feel for each other, and all their manipulations won't make us lose our principles. Okay?"

"Frall, you know what I think?"

"What?"

"That you're the most wonderful guy in the world?"

"Which world?" he asked, laughing, as he kissed me.

"All of them."

Of course, that caused more kisses, which slid into whispered words of love, but then we both backed up, knowing where that would lead. Besides, there'd be no fire or dinner if we didn't get going.

I'd never gone hunting before. It had always been the men who had headed out into the forest, leaving Franca and me behind. Frall was very proud of his accumulated knowledge and eager to share it with me. He showed me how to create a trap, explaining everything like a teacher, step by step.

First, we had to weave the vines. Frall showed me how to loop it, and we placed that end on the trail hidden by pine needles and dirt. We hid behind a bush and waited. Frall had looped the other end of the vine over a tree and was ready to pull the loop closed on any unsuspecting prey. I was pretty sure that no animal would be that stupid. Hadn't we put our scent all over the vine trap? Of course, I didn't say anything, so we sat waiting, staring at each other.

I was wrong. The first fuzzy critter that walked down that path stepped right into the loop. Frall pulled, and the thing became a squealing bundle of teeth and nails. I didn't want Frall to kill it. On that, we argued, but he had the knife, the one he always carried, and despite our disagreement, he slashed at the animal's throat, and that was that.

I felt sick as I saw the blood. I turned away, not able to look at it.

"Clista, if we're here in the forest long, you'll have to adapt to killing animals. You understand that, don't you?"

I didn't argue with him, but I still didn't look. The animal had been *cute*. Frall, not looking at me, untangled the vine, curled it up around his hand, and tossed the dead animal over his shoulder.

I didn't mind collecting firewood, and I was glad when Frall showed me how to make the fire. That was fine, but I didn't want to learn how to skin a dead animal. That part I had no intention of learning. Frall took one look at my face and shrugged. I guess he figured there was no way he would change my mind — at least, not that day, anyway.

Frall hunted around for a suitable spit to roast the meat on. The beastie was soon cooking and fragrancing the air with the delicious smell that meat has to entice the hungry. I was starving by then, but the thought of the poor, dead animal still turned my stomach. Even when it was ready, I refused to eat it.

"Good. That means more for me," Frall said, and he tore off a chunk with his knife. He tossed it back and forth between his hands because it was still too hot to eat or to hold. Because I was laughing at his show, he laid the meat down on a big stone to let it cool.

"You're going to get hungry," he warned.

I shrugged. "Maybe we'll find the same nuts that Smog located that time."

"I hope so. I don't like mornings without coffee."

Frall tore a piece off the slightly cooled chunk and popped it into his mouth. "Wow!" he said. "This is stupendous!"

I turned away and stared up at the tree, the "grandfather tree" I'd dubbed it, much to Frall's delight.

I was still staring up into the redwood's height when Frall attacked. He launched himself at me without a word, and his fingers began to tickle. I fought him off, laughing, rolling about on the soft but prickly, padded ground. When at last I called "uncle," Frall stuffed a piece of the meat into my mouth. He'd had it in his hand the whole time.

I started to spit it out, but I had to admit, it was delicious.

"Darn you, Frall," I said, and he sat up grinning the same way that Smog always did when he'd told one of his strange jokes. Had Frall copied Smog, or was this just the grin of a man who has won his battle?

After that, it seemed silly to pretend that I didn't want some meat. Frall handed me a piece, and I took it. Then, both of us were gnawing at it like cave dwellers. Later, we wrapped the leftover pieces in a torn-off strip of his shirt and placed it at the side of the fire. "Breakfast," Frall said, smiling at me.

Forests have fewer hours of daylight than open expanses do. We huddled together, sitting by the fire, toasting our front and then our back, laughing a bit as we changed positions, still talking the whole time. Then, when the darkness was full, we cozied down in an area we'd raked with shoes and sticks. Even so, the ground was still a bit scratchy. The needles were prickly, but we had no blankets or extra clothes to lay beneath us. Our bodies and the fire were our only warmth.

We spent the evening hugging each other and trying to stay warm. Once again, I felt Frall's breath on my neck and his arms around me. For some reason, that made me feel unbelievably content. I closed my eyes, and listening to the cracking of the fire and the monkeybirds that called to each for a while longer, I fell asleep.

Through the night, a pair of jasmire watched us. Frall woke me to point them out. The cats' shiny green eyes were glowing in the darkness, but they didn't approach. We stared at them, worrying about their presence. We whispered countermeasures, put more wood on the fire, and prepared for the jasmire's attack, but it never came. Finally, sleep overcame us, and we slept until morning.

Chapter Eight

I woke with Frall's loving eyes staring into mine. It was a good way to start the day, almost as good as a cup of coffee, something we, didn't have.

"Good morning," I greeted Frall. He smiled, stood up, and stretched his limbs until they cracked.

"I'm supposing it's a great day to go fishing," he suggested. "What do you think?"

I laughed and admitted that I had no idea how to fish. Lessons began right after we each slipped behind a bush and gave our water to the soil.

We carried our packet of meat down to the brook, searched about for a couple of "poles," and upon finding two good ones, discovered that we had nothing to use for hooks.

"We need something sharp and metal," Frall said, glancing around the area. Of course, there was no metal lying around. Frall searched his pockets. We both inspected our shoes. Then we sat down by the stream and nibbled on the leftover meat, trying to decide how to make a hook.

"What did the first people use for fish hooks?" I asked.

Frall shrugged. “Maybe they caught fish with nets. I heard once that the long ago Earth Native Americans could catch fish with their hands, but I’ve tried that, and it doesn’t work — not for me, anyway.”

“With their hands? Eww,” I grimaced at the thought. But the more I started thinking about it, the easier that sounded. “Give me your shirt, and I’ll try it,” I said.

“Why don’t we use yours?” Frall asked, grinning at me, just to see if I’d blush. I didn’t. In fact, I took no notice of his suggestion, despite the way Frall was hooting and snorting through his nose like some kind of strange bird. Still, without further comments, he handed me his shirt. By that time, he was only chuckling a little.

I removed my shoes and socks, tied my long dress into a high knot, ignored Frall’s whistle, twisted the fabric of his shirt around my hands, and waded in. “Whoa! It’s cold,” I complained.

Frall leaned back contentedly against a large, warm boulder, looking far more comfortable than I was. I waded in deeper, shivering The good news was that the creek was full of fish. I could see fairly large ones swimming about. They were brazen, too, flitting between my legs as if my white, pale limbs were the posts of a bridge they needed to pass under. I cupped my hand and stabbed at a couple of them, but it came up empty.

Frall’s laughter irritated.

“I can do this,” I said. “I’m just warming up.”

“Sure,” he teased. “Did you learn how to catch fish in Leadership or in that fancy math class you were taking?”

“Be nice, or I won’t share my catch,” I warned him.

That sent him into extremely loud-mouthed hysterics. He practically spasmed with laughter.

"All right, here comes a big one," I said. "Listen, fishy, how would you like to be breakfast?"

I swooped down on the fish. I thought I had it, but it slipped through my net of material as if it was playing keep away.

"How did he do that?" I roared, kicking at the fish with my foot. Of course, I didn't come anywhere near him, but I did splash myself and Frall.

"Hey," he complained, standing up and backing away. "I already donated a shirt to this business. Don't get my pants wet, too."

I suppose I could have mentioned the way the bottom of my dress was getting soaked, but I didn't. I had my sight on another fat fish. I was tracking him. The moment he came into my bridge area, I dived in on him. I ended up wet as the fish, and the fish swam away, not even pausing to act scared.

"Are you okay?" Frall said, taking off his shoes, probably so he could wade in and save me.

"Go away," I told him. "I'm not done yet. I know I can do this. It's just a matter of convincing the fish. Don't stir up the water."

I stood up and tried again, but the drips coming off my dress kept disturbing the image of the water. The next time I grabbed, I was even further from success.

"The water distorts your projection, you know," Frall told me, no longer laughing.

"I can see that," I told him, "but I know I can get this to work."

Frall took his seat on the boulder again, his legs crisscrossed. "Carry on," he said. "I'm certainly enjoying the show."

That made me look down. My dress was clinging pretty badly. I pulled it away from my body, ignoring the heat in my face that warned of my blush. "Look somewhere else," I demanded.

Frall closed his eyes and pretended to sleep.

Meanwhile, I spotted a fish bigger than the others. It was coming right at me. I let out the material of my "net" for a larger expanse. Then, slowly, I pushed it into the water, holding it still as if it were fish shade. That lovely fish moseyed over to me. I think he was thinking about taking a nap, just like Frall. He never knew what hit him.

"Got him!" I cried out, bringing the thing up out of the water. But the darn fish was heavier and floppier than I'd expected. "Frall, help me. The fish is putting up an awful fight. He's trying to jump out.

Luckily, Frall still had his shoes off. He jumped in and plowed his way through the water. "I can't believe you did it!"

The fish was still waging a flipflop battle. It took both of us to land it. Once ashore, Frall took out his knife and stopped the poor fish's struggle.

"Oh!" I cried out, turning away. "I forgot about that part."

"You're sure not much of a hunter if you can't kill your prey. Besides, it's easier on the fish than letting it die from lack of air.

I peeked, trying not to remember how beautiful the fish had been swimming in the water.

"Tell you what," Frall said, probably seeing my green-tinged skin. "It's only fair. You caught the fish. I'll skin it."

While Frall was making good on his offer, I laid his shirt on the boulder so it could dry. The sky, the parts of it I could see through the expanse of tree limbs, was still pretty overcast, so there wasn't much sunshine, but the boulder was warm. I figured Frall's shirt would dry there better than just hanging from a tree limb.

I wished I could take off my dress, too, since it was sopping wet, but I had nothing to change into. "I'll meet you back at the fire," I called out, running up the slight incline of the stream's bank.

Frall didn't say anything, but he grabbed his wet shirt off the boulder, moved the fish back onto it, and followed me.

"It isn't a good idea to leave my sight, Clista," he scolded when we reached the campsite. You know I'm the only one with a knife."

"But I'm freezing," I told him, my teeth chattering a drumbeat.

"I'll get more firewood," I told him, "and I'll just be over there, right in sight, okay?"

He nodded, and without glancing up, proceeded to prepare the fish for cooking.

One of the reasons we'd chosen this particular spot for our campsite was its great source of firewood. A tree had fallen and was lending itself to easy pickups of chunks and pieces of logs. I grabbed up an armload and lugged them back, letting them drop onto the ground beside the fire.

"You know you need to take off your dress," Frall said, probably observing that I was starting to turn blue.

I glanced at him. "No way."

"You're wearing stuff underneath, aren't you? Isn't that the same as a bathing suit?" he asked.

"No."

"Look, I'd offer you my shirt, but you already got it wet . . . "

"I'll be fine."

"If you get sick, what will we do then?"

"They'd probably bring us back."

"Are you sure of that?"

He stopped to look at me. I shrugged. "All right. I'm not sure of anything — not when it concerns the *Old Ones*."

He nodded and continued cutting. "Clista, I'm really serious. I want you to take off the dress and dry it. I promise not to look. I'd leave, but I can't do that. It's not safe."

"It wouldn't do me any good, Frall. Everything underneath it is wet."

"Yeah, but it would get most of the wet off you. Do it, Clista. You're shivering worse than when you got out of the water."

"I can't."

Why couldn't he understand that? Some things just weren't right, and this was one of them.

For a moment he didn't say anything, but I could feel the steam rising of him as he continued being irked with my stubbornness.

"Remember how I got you to eat the meat?" he said, breaking the silence.

I remembered. He'd pounced on me and wrestled me into acquiescence by tickling. "You wouldn't do that . . ." I whispered, backing away as if there were any place to retreat. But Frall was looking at me with an expression I'd come to know, one that gave me no doubts that he'd do exactly that if I didn't comply.

"I already gave you my word, Clista," he added. "You either trust me, or you don't."

My dress was a clinging, soggy mess. I could barely even get it off, but I did finally. Conscientiously, avoiding even looking in Frall's direction, I dragged a large tree limb over to the fire and stretched my dripping dress over it. Then, hugging myself for warmth, I sat as close to the fire as I could comfortably get.

Frall stood up and came toward me.

"What are you doing?" I cried out, panicking.

"I'm not going back on my promise if that's what you're worried about. I'm just putting the fish onto this thing I made."

So saying, he held a vine-tied twig platter up for me to examine.

"Smog used something on the material to keep it from burning," I said.

"Yeah, you have any of that with you?" Frall asked.

"Wait," I said, examining it. "This is just going to burn up the moment you put it into the fire. It's not going to work."

"Yeah, tell me about it. What would you suggest?"

I turned to the side and let another part of me get warm. While shifting about, I stared into the flames and thought about the problem.

"What if you made two of those and put the fish in the middle? Then, if you could attach a handle, you could move it in and out of the flame whenever it got too hot."

Frall stared into space for a moment, thinking about it. Then his smile broke out. "Hey, that's not a bad idea. Good going, Clista. How would I get a handle on it?"

While Frall was looking for more twigs and vines, I thought about that one, too. So did he. When he returned from harvesting the fallen tree branches, he had the answer.

"Look, it's simple, Clista. I'll just weave in a whole branch. It can be like one of the twigs — part of it, you know? See what I mean?"

I nodded happily. It was fun working together like this, each of us adding something. Wouldn't the proctor be proud of us?

"That's perfect," I said. "Need any help?"

He shook his head. "Nope. It's better if you stay where you are, I think. I'm doing the best I can, Clista, but this situation isn't getting any easier for me."

"You're saying I should put my dress back on?"

"No. Just stay where you are."

I couldn't exactly do that. The fire was burning my right side. I turned my back to it and tried to cover up as best I could with two bare arms and humped shoulders.

"Remember how we used to meet before school, Clista?" Frall asked out of the blue.

I looked over at him and smiled. Frall was staring at me. I glanced at my dress. It had stopped dripping, at least.

"Clista? Do you remember how simple things were then? I used to get so upset about a silly test or a report I had to give. And I was worried about my future, worried that the wrong company would want me."

I laughed. Frall was sitting on the ground weaving his fish cooker. I couldn't help noticing his broad shoulders and the way his back rippled with the lean of his muscles. I sighed.

“Yeah. It’s strange how everything changed so quickly.” I said as I shifted to allow another part of me to get warm.

“Some of it was good, though,” he added.

“Like what?”

“Like us, Clista. Would there even have been an “us” if this all hadn’t happened?”

I knew what he meant. We’d both been on track for moving into a company training program. We wouldn’t have had time for relationships then. Would we have continued our friendship? Probably not. Training programs often sent employees to one of the moons or at least to the other side of Rwawan. The odds are that our lives would have shifted in completely different directions.

“I doubt it,” I answered honestly. “But remember the proctor who was on my back? Wouldn’t he be happy seeing us working together?”

Frall laughed. “Yeah, I can’t imagine his face, seeing us collaborating. But don’t forget, he’s the one who set us on opposing sides. He instigated our rivalry.”

For a moment, we were both thinking back to that week. We’d broken down all those barriers and turned into really good friends . . . and more.

“Strange. Whichever company had hired us, we’d have been gobbled up into their family. That’s the way they did things,” Frall added.

I nodded and shifted my body again from the fire’s intensity. My toes were burning, so I had to withdrew them from the heat, even though, I knew they’d soon get cold again.

“We won’t be able to go back to our lives there, Clista. You know that, don’t you? No company would hire us — not disappearing, as we did for a year. And neither of us got to finish our last year at the Academy. Do you suppose we could ever make that up?"

With everything going on, I hadn't had time to think about how we'd completely dropped out of our former lives. How stupid of me. Frall was right. Someone chasing you, trying to kill you, or at least all your friends and family — that shifted your priorities. But Frall was right. We couldn't go back. Probably not ever.

"Even so, it was all worth it. I mean that, Clista."

I reached over and turned over my dress. It was still wet but warm from the fire. It was starting to dry. By nighttime, I'd probably be able to wear it again. I glanced at Frall. "I don't know what to say. I mean, I'm glad of that, too, that there's an *us*, but if you mean was it worth it that our parents…"

"Of course not. No, I didn't mean that. Losing our parents was the worst. It hurts and will continue to hurt for the rest of our lives. It's so final. No goodbye. No thank you for raising me, for putting up with . . . well, all the stuff that kids do. You know."

I did. I felt guilty for the way things had been left, too. I guess there are no perfect kids — ones always ready to clean their rooms, scrub the bathrooms, and never whine or talk back, not that I'd done that very often. But I wished I'd told them *I love you* more and, like Frall mentioned, had been given a chance to say *thank you*.

'I just meant that you're the one good thing that's come out of all this," Frall said.

"I feel the same, and thank you," I murmured, staring down at my feet. I'd just put my shoes back on. It probably looked silly wearing them when I had my dress off. But the pine needles underfoot were sharp. Going barefoot was painful, and I needed to use the bushes for private matters.

I started to say something, but I got tongue tied. Frall's praise had made me uncomfortable. Speaking of uncomfortable, I suddenly realized that my skin was burning. I rotated again.

"How's the new fish cooker doing?" I asked, avoiding the subject.

"I just finished it. What do you think? Should I put the fish into it?"

The contraption looked ready. Whether it would fall apart in the fire was something else. "Try it," I said. "You want me to hold it?"

"No. I'll do it. Thanks."

Frall loaded the fish into its holder and walked over to the fire. "All right," he said. "Cross your fingers."

The cooking apparatus held the fish nicely. Frall lowered it a bit, keeping it just slightly off the flames. At first, the vines curled up, crinkling their outer leaves, but it held. Frall looked really uncertain, his face full of doubt.

"You're so good at everything," I told him. "You know how to trap and fish and cook. You're a really great person to be dropped into the forest with."

Frall smiled. "Think so?" He laughed, but I could tell he was glowing from more than the heat.

"What do you want to do once your dress is dry?" he asked.

"Go home?"

"That's easy," he said, laughing again. "Just click your feet together and say, *There's no place like home*."

"Very funny."

I excused myself for my private time with the bushes. It wasn't fun to be back in the wilderness for that and the leaf wipes did a lousy job.

A quick sprint down to the stream allowed for a hand washing, but we hadn't yet found any flowers that acted like soap. I used sand, but I doubted it did much to get a thorough cleaning.

When I returned, shivering again because although I was dry, a breeze had started up and was sending chills. But the fire was crackling, its heat a welcome warmth. The smell of the fish was already a mouth-watering enticement. I wondered how I could be so hungry again.

"I know. After lunch, let's go see if we can find some nuts. Okay?"

"We need something to heat the 'coffee' in, too," Frall said.

"Shoot. You're right. We used a hollowed out tree trunk before, remember? But it took forever to cure, and it had bugs in it. Guess we'll just have to start all over again and make one. I wish we knew how long they were going to keep us here."

"Yeah, me, too," Frall said, turning the fish over. "You know, I was really impressed with your fishing ability. I never could have caught this fish. Only I think we have to find a way to fish that doesn't get you wet."

"That would be nice," I laughed, moving again to cover another side of me with heat.

The fish was delicious. We ate all we could hold, and then Frall wrapped the remainder in the same piece of his shirt he'd used for the meat.

Then he walked over to check on my dress. "It's almost dry. You want to put it on, and we can go check on those nuts?"

I froze, the shivers dancing up and down my spine, even while my back was soaking up the fire's heat. "We've got a visitor again, Frall," I said, eyeing the thicket.

"Darn it. Must have been the smell of the fish."

"Go catch your own," Frall yelled at the jasmire.

If it was about to attack, I wanted to be fully clothed. I slipped my dress back on, enjoying the warmth, even though it wasn't quite dry. The heat wouldn't last, but it felt luxurious for the moment. "You were right, Frall, about drying the dress. It feels much, much better."

Frall was tending the fire, tamping it down so it would still be burning when we returned. I guess he was still planning on us looking for those nuts, but what if the jasmine decided to attack us? Should we be banking the fire? Wouldn't it be better to build a huge bonfire to keep the animal away?

"Yeah, your being dressless just about killed me," Frall said, moving closer. In a moment, his arms were around me, and his lips touched mine. "You have the most incredible body," he whispered into my ear.

"I thought you promised not to look," I complained.

His arms tightened. "I tried, Clista. Believe me, I tried."

It was only for a moment that we luxuriated in the embrace. Then, without a word, Frall released me and turned away. "Come on," he said, taking my hand. "We'd better grab a couple of sharp branches just in case our friend decides to get too friendly."

We hiked deeper into the forest, marking our way with an occasional knife cut on the trunk of a tree, weaving our way toward the north — at least, that's the direction Frall said we were going because of the pattern of lichen growth.

As we walked, we held hands, but we talked only sparsely. We were both listening to the jasmire's padded paws on the soft forest floor behind us.

"It's definitely following us," I said.

"Yeah, I know," Frall whispered. "I can't understand why. There must be easier game for a big cat like that. But what do I know about predators on this planet? At least, I think we're still on an alien planet. Maybe we're back on Rwawan."

That stopped me. I looked around. How could anyone tell? It sure didn't look like the area around the ship or the palace where we were staying on Flagan Pren.

"When the jasmire decides to charge, you stay behind me, okay? I've got the knife."

"When a jasmire attacked before, I used magic. Do you think the *Old Ones* will help us again?"

Frall stopped and turned to stare down at me. "Wait a minute. Help you? Don't you remember what they keep telling you? They don't have to do anything. *You* have the magic. You and I together."

I laughed, a sickly, scared kind of chuckle. "I hope the jasmire knows that. Something tells me it may not believe in magic."

Frall pushed me up against a tree. "It's coming, Clista. Keep your back solid against this trunk. I'll hold it off if I can, but if it gets past me, climb up the tree, you hear me?"

Frall's new strength was overpowering at times. His push almost knocked the air out of me. I stood, gasping for breath, not able to argue for the moment.

The jasmire had come into plain sight. Teeth longer and sharper than Frall's knife, a coat heavy and golden and slightly textured with shag. Her eyes glowed the green of a brand new leaf. If I'd had the breath, I would have gasped in appreciation.

I wheezed in air and stared into her eyes. "Nagarth," I whispered. "Nagarth."

"What is that?" Frall asked, not turning to look at me.

The cat's eyes were on me. She was ignoring Frall.

"Nagarth," I said again. "It means sister."

The cat took another step forward and then leisurely sat down, her tail switching nervously like a side-winding snake.

"It's okay," I told Frall. "She's not hunting us. She wants to be friends."

"Clista, have you gone crazy?"

Frall didn't believe me, although I tried to explain. He took a step forward and made a sudden lunge at the cat.

"No, Frall. Stop," I cried.

The cat's huge eyes turned to stare at him. She blinked and then focused, turning her stare full force onto Frall.

"Stars!" Frall said, halting his attack. "You're right, Clista."

"Nagarth," Frall said, greeting the cat and taking another step toward her, but his knife, just in case, was ready. "Nagarth," he said again. "That's the *Old One's* language. Where did we learn it?" Frall mused.

"From the sleep tapes, probably," I told him as I watched the jasmire, admiring the fine black lines of her beautifully decorated face. Her face was exquisitely rimmed with streaks heading out in all directions diagonally. Her nose, a triangular pink, was surrounded by the cutest little dots. Her whiskers were shiny with black. She was a gorgeous animal.

"Buff," she said, glancing at me as if to thank me for my thoughts.

Frall had continued forward and was almost a foot away from her. He stretched out his hand and held it near her muzzle. The jasmire reached out and met his hand, breathing in Frall's scent.

"Buff," she said again. "Buff."

The jasmire was not a small animal. In fact, she was almost as large as I was. Her teeth and claws made her a formidable enemy, but this one didn't seem interested in killing. She lay down and turned her belly to Frall for a good rub.

"This is too weird," he said.

I watched as he stepped forward and then bent down, his hand stroking the jasmire's belly.

"She's purring, Clista."

Another jasmire poked its head through the bushes. I swiveled to look at it. "Frall, back away. There's another one, and it might not be happy about you touching its mate, if that's what she is."

Frall looked up. Quietly, he backed away. "Sorry, kitty. I'd love to pet you, but it looks like we've got company."

"Buff?" asked the first one, still stretching paws that were retreating and advancing in pleasure at the attention.

The second cat came out of hiding. It eyed us and growled. Then it walked over to its mate, licked at the cat's face, and stood, staring at us.

"Nagarth," I said softly. The eyes of the new jasmire slid to me. They were intelligent. Its ears went forward, and it studied me as if waiting for more words.

"Nagarth, tagone blapote," I told it.

"Wow!" Frall exclaimed. "I know exactly what you're saying. It's the *Old Ones'* language. *Friend, no harm intended*. This is so cool."

The second cat put its ears back at Frall's words. It growled and took a step forward, its claws descending to full attack.

"Nagarth. Nagarth," I repeated. "Tagone blapote."

Once more, the cat swung its attention to me. "Buff?" it asked. "Buff?"

"Stay where you are, Frall. This time, it's my turn," I said, walking forward.

Frall grabbed at me. "No way, Clista. Get back against the tree. This guy is not friendly."

"Stop telling me what to do and let go of me before the jasmire attacks you. Can't you hear it growling?"

"Yes, and that's why I'm not letting you go. Get behind me."

I recognized the stubborn look on Frall's face. No words could talk him out of it. I tried to remember what they'd taught us in Leadership class and in Anger Management about such situations, but neither provided answers. Frall was overstepping his boundaries — way beyond.

"I mean it, Frall. Let go of me," I said again.

"No. Get back against the tree. I'll take care of this, Clista."

With my heel at a sharp angle, I stamped. Frall let out an unmanly "ow!" and grabbed at me again, but I'd used the moment of surprise to gain my freedom. I walked toward the huge cat, my hand level and open.

"Darn it, Clista. Don't be so difficult." Frall lurched towards me. I dodged his hand and took three more steps toward the second jasmire.

That jasmire was growling louder, but his eyes were focused on Frall, not on me. I took another step. "Nagarth," I told the animal.

The cat's ears flicked forward. He leaned toward me, his nose touching and breathing in my scent. Then his huge pink tongue licked the skin of my palm, and I swear he smiled.

"See, I told you so," I said over my shoulder.

"We need to talk about this," Frall snarled, in a voice almost as threatening as the jasmire's initial growl. Frall took a step closer to me, as if he still intended to grip my shoulder and jerk me away.

The jasmire's huge mouth opened, and he roared a warning at Frall.

"Clista, get away from it! Can't you see it's going to attack?"

I ignored Frall and plopped down next to both animals. The big male tongued my cheek and purred into my ear. The female once again exposed her belly, begging for a scratch. I was happy to oblige her with my left hand, but my right one slid onto the male's head, stroking him, hearing his purr grow into a loud rumble.

"Stars, Clista, but you're the most difficult woman I've ever known."

I laughed softly. "Only when you try to muscle-man me."

"Is that what you think I did?" Frall took another step. The male jasmire didn't roar again, but he watched Frall, his ears sliding back and forth, his tail flapping about like a wounded bird.

"Be careful. Merot doesn't like you very much," I warned Frall.

"Merot? You've named him?"

"He told me his name. Weren't you listening?" I said, sliding my hand under Merot's well-muscled jaw.

The animal's purr started up again. He leaned his weight against me, curling his tail around my waist.

"I wonder if it thinks you're part of his harem?" Frall suggested.

"Don't be disgusting," I snapped. "Why don't you just talk to the cat? There's no reason you two should be enemies."

"Get real, Clista. That animal doesn't want to talk to me. It has exactly what it wants, your attention."

Merot shifted away from me and began to growl deep in its throat. His fur ruffled, creating a coat a full inch taller and bigger all around.

"Merot, blagonel pote."

Merot turned his huge, green eyes on me, studied me for a moment, then brushed his head against my shoulder. Once more, his tongue sandpapered my cheek.

"I think it's obvious you prefer his kisses to mine," Frall kidded me, but I felt the ring of bitterness in his voice.

"Merot, blagono pote. Nagarth. Se cope ano," Frall said.

"I belong to you? Since when?"

"Be quiet, Clista. I'm attempting to make friends with the animal. I'm just using terms a cat can understand."

"Nocopo al," I said, meaning *I am not his*.

Merot's eyes regarded Frall and then me. He yawned hugely and lay down beside me. "Buff," he said, meaning, I think, that Frall was permitted to come forward.

"You didn't have to say that," Frall protested. "What if it thinks you aren't mine? What if it thinks you're available?"

"Stars, Frall. He's a jasmire. Why would he care? Just come and make peace with him. Both jasmires are incredibly soft. I love the way their fur feels."

"Yeah, I like the way yours does, too," Frall said teasingly.

I put my arms around the male's still upright head. He purred at me, still watching Frall as my boyfriend walked slowly closer, indicating with his hand that he wanted to be friends.

"Nagarth," Frall said.

The huge cat sniffed and then sniffed again. His ears flapped back, and he growled low. I threw my arms around its neck. "Nagarth, nagarth," I said.

Merot obviously didn't know what to do. His tongue shot out at the side as if rewarding me, but it didn't quite make contact. Then Merot rose up, shaking me off.

"Frall," I warned.

But Frall was standing his ground. "Se cope ano," Frall declared again, once more telling the cat that "the female belongs to me."

I glared at him. It was so ridiculous. Was Frall jealous of a cat?

Merot turned and looked me in the eye. He roared and then sat down. He didn't have to issue his command so loudly. I understood with the force of his thought alone.

Frall's hand was armed with a knife again.

"Are you mad? Don't you dare hurt him," I cried out.

Frall ignored me. The two of them were staring into each other's faces and both of them were snarling.

"RRRROW," said the female, suddenly getting up. Her ears were back, and she repeated her protest several times. I reached out to touch her, but she shied away from me.

"Trepok trezet, Trepok trezot, Trepok trezit, Trepok trezut…" I began the chant we'd used the first time in the forest to connect to the *Old Ones*.

Both cats swung their heads to look at me. "Claydeah," I told them.

The large one roared at me, but the smaller one came and sat back down. She rubbed her head against mine and placed it under my hand so I could pet her.

"Claydeah," I said once more in a demand for obedience.

Merot roared, but it was half-hearted. Then he, too, returned and sat at my side. We all watched Frall.

"All right, the forces are lined up against me here. What do you want from me, Clista?"

"I'm your mate, Frall. I accept that, but I'm not less than you. You can't tell me not to go forward when I think it's the right thing to do or protect me all the time — not when I don't want or need it."

Frall groaned. "You think that I believe you're less than me? Stars, Clista. You're always amazing me. You're incredible in every way. Why would you believe I thought that, even for a moment?"

I didn't answer. I thought he hadn't heard the rest of it. I was wrong.

"I can't help it, Clista. I love you. I don't want you hurt. Can't you understand that?"

He was still keeping an eye on Merot. The knife in his hand was horizontally ready. "Put away the knife," I told him.

"So I'm supposed to take orders from you now?"

I ignored him. "There will be no more: "Se cope ano, the female belongs to me."

Frall stared at me a moment. "The cats aren't going to attack?"

"No. You can put away your knife."

"All right, Clista. I believe you, but you can't have it all your way. "Clista cope ano fe Frall cope ase."

"Clista belongs to me, and Frall belongs to the female." I could accept that. It was fair. I nodded.

"Walk slowly, Frall. If the female accepts you again, Merot will, also."

"What is her name?" Frall asked, stepping closer.

"She is called Yte."

And that is how Merot and Yte accepted us into their world. From that day, they stayed close, frolicking and hunting with us. Merot caught fish much better than I did, and he shared equally with us. In fact, after he'd tasted our cooked meat, he gave his entire kill over to the fire, and we feasted like four good friends, sharing and savoring the meat equally.

Yte and Frall quickly became the hunters. Yte honed Frall's hunting skills, and together, they started bringing down big animals. Frall and I soon had new coats and blankets for our bed.

At night, instead of watching from the bushes, the jasmires slept at our sides. Our nights were much warmer and our days much safer with their companionship.

Sometimes though, late at night, the cats would leave us and go prowl the forest. Then Frall would hold me in his arms, and we would whisper things we didn't say when the cats were around. Then our lips joined and mingled, and our hands frolicked, but we were always careful. Promises kept us from romantically journeying too far. Personally, I think it made our relationship sweeter, but there were times that Frall cursed and sat away from me. Then, as I watched him, I grew less certain of resolutions. Those times were difficult.

Chapter Nine

One day, when Yte and Frall had gone hunting, and I was alone with Merot, my monthly period came. I'd been expecting it and had prepared. I tied the rags and matted leaves to my body. Then I went down to the water and tried to drive the fish into Merot's fierce teeth. But cramps overcame me, and I was forced to go back to the fire to lie down.

Merot whined, but he seemed to understand that I wasn't feeling well. At last, he left me and returned to catch our fish. I think I was half asleep when I felt the presence of another jasmire, one I didn't know. It crept up behind me. When I woke, it was baring its teeth in a low, hunkered position, one that meant it was about to attack.

"Merot," I called, but I knew there was no time for his arrival. "*Nagarth*," I tried, but the huge cat slinked forward, still low to the ground and readying itself for a fatal blow as it sniffed the air and grinned.

"*Nagarth*," I said once more and touched my necklace, seeking the power, or at least the confidence it gave me. It warmed to my touch, but the cat didn't react to it. The jasmire edged still closer, its haunches preparing.

I reached over and pulled a branch from the fire and held it in front of me. When the animal leaped, I screamed and thrust the branch at its chest. That stopped the attacker for a second, but then it seized the stick in its strong jaws and jerked the branch out of my hand. The animal spat the branch to the side and whirled around to bite before I could reach for another.

"Frall," I screamed, but I knew that was useless, too. The jasmire, ignoring the supposed magic of the *Old Ones'* protection, gave a final roar. Then it attacked. Its heavy body knocked me down. Its teeth sliced into my arm. I screamed, but that was all I could do. I couldn't get up. I couldn't get free of the dreadful teeth and the animal's firm grip as it tore into me.

Pain hit me like fire sticks rammed down on my naked skin. I fought. I cried. I screamed.

"Nagarth, nagarth," I kept saying, but the jasmire wasn't listening. I babbled magic words, but none of them worked. The sharp teeth were like stinging nettles.

And then Merot was suddenly there, a hurling ball of fur. He rammed the strange jasmire, toppling it into the fire. The stranger howled in pain and flung itself into the dirt, where it rolled until the fire was out. Merot attacked again. Then the fight went on. But I think the fire had cured the strange cat's bloodlust. As soon as Merot permitted, it bolted and ran, its tail between its legs.

There was blood all over me. I attempted to stem it with scraps of Frall's shirt, but Merot whacked me with his body almost as roughly as he'd done the invader. Then, towering over me, he ripped off the wrapping I'd placed and began licking the bites.

"Stop it," I told him. "Let me up."

But Merot wouldn't listen. He lay down on top of me, an immovable object, and continued to lap at my blood.

His tongue stung. It was like rubbing burned skin. I cried and begged him to stop, but there was no way I could wiggle out from underneath him. He nipped me gently each time I tried.

I knew then that Merot was going to kill me. Wasn't that the way of the wild —kill the wounded? He, like the other jasmire, would pierce my flesh in a minute, but for now, his greed was solely for my dripping blood. It should have mattered to me, but it didn't. I was too weak to care. And when the pain of that rough-textured tongue licking at my wounds grew too much, I simply gave into it and faded out.

I slipped in and out of consciousness for a while. Sometimes, Merot was licking my face. Other times, he was still tonguing my arm. The pain was fierce. I couldn't stay awake for more than a minute. "Frall," I'd whisper, but he still wasn't there, so I'd close my eyes and sigh, then drift away again.

At one point, Frall was finally there beside me. But nothing had changed. He allowed Merot to continue eating me. I couldn't understand that. Why would Frall turn on me, too? Why did he allow Merot to torture me?

"Get him off of me," I cried more than once, but Frall held my hands to keep me from pushing the cat away. It was obvious he had stopped loving me. He had struck a bargain with the jasmires. I was being sacrificed to the cats.

But then I woke, and the agony was more fierce. Strangely, I felt stronger. Frall was there beside me, Yte was supping on my arm.

"Why do you torture me? Please, let me go," I screamed, once more feeling almost strong enough to fight for my life.

"I can't, Clista. Yte's working out the poison. You've been burning up from it. She's trying to save your life."

Frall was bathing my face. I wanted a drink, but I'd already given all my energy with my struggles to be free. I didn't speak. I couldn't. I simply closed my eyes and let unconsciousness take me back into peace.

When I woke again, it was morning. Yte was gone, and Frall lay at my side, his arm sprawled across me. I tried to move it. I wanted to crawl down to the stream.

The moment I attempted to move away from him, Frall woke up. His hand shifted to feel my forehead.

"Stars," he cried. "You're still burning up. I'll get water. Don't move."

"The necklace didn't save me," I told Frall.

"I know," he said as he returned with a carved-out wooden cup of water. "Sip this."

He lifted me up, and I cried out. My arm must have fallen into the fire. It burned. Maybe coals were eating away at my skin.

"I'm on fire!" I screamed, sobbing from the pain.

Yte, like a speeding pilotcar, knocked me back down. Once more, she attacked me. I wept as she drank my blood. The fire in my arm flamed higher.

"Why?" I cried out. "Why?"

Frall bathed my forehead again. I tried to ask him why he allowed the cats to hurt me so much, but I couldn't. My lips were parched. My throat was so dry it was as if I'd swallowed ashes. Perhaps Frall was punishing me. I'd often been mean to him. It made perfect sense he would want his revenge.

"Frall," I cried.

"Hush. Sleep, Clista. It will be better tomorrow."

"Buff." Merot was back. Yte left. Merot came and lay down beside me.

"Why?" I asked, but no one answered. The pain raged. Merot's tongue continually fanned the flames. I passed out again, my body an inferno of agony.

When I woke, Frall was stripping off my clothes. Once, I would have fought him, but I lay placidly, not bothering to argue. It didn't matter anymore. Nothing mattered. I was dying.

Frall picked me up and carried me down to the water. Even jostling my arm as he did sent me into unconsciousness. I woke to find myself immersed. Frall was in the water with me, holding me up.

"Why?" I asked again.

"I'm cooling you down," he said. "Drink some," he said, his hand cupping the liquid.

I drank a few drops, but I didn't want it. I only wanted to sleep. I was finally cool enough. The fire was dying down.

Unfortunately, Frall did not leave me in the water where I wanted to stay. He lifted me up again and carried me back to the camp. Then he threw a fur blanket across me.

"No," I cried out and tossed it off of me.

But Frall demanded, and he was stronger. The fur covered me once more. Again, I burned.

Yte came over and made it worse by drinking my blood again.

"Why?" I asked Frall, but he turned away and didn't answer.

I woke each time one of the cats came to lick me. I hated Frall for allowing it. It went on and on. Sometimes Frall wasn't there, but other times he watched. Those were the times it hurt the most. Betrayal burns as much as fire sticks.

And then I woke, and I was not aflame. In fact, I felt chilled. My teeth were chattering. I searched for the fur, but it wasn't nearby. I tried to curl up into a smaller body, but my arm was lead. It wouldn't move.

"Frall," I cried. He was suddenly beside me. His body leaned over mine.

"I'm so cold," I said, shaking like a wind-blown snowflake.

"Good," he said. "Thank the stars. I'll get the fur, my darling."

I cuddled into it when Frall placed it across my body, but as I curled up, my arm once more blazed with pain, and it was stiff.

"What's wrong with my arm? I can't move it," I whined.

"You were thrashing about. You kept tearing it open. I had to tie it to a heavy fallen limb."

My arm was a throbbing knife at my side. It ached from being tied. It burned from the cuts, and I could feel that it was swollen with infection. I tried to turn and look at it, but Frall prevented me.

"Here, I want you to drink this," he said. He lifted me slightly and held my head, then placed a wooden cup at my lips. "Drink it all if you can. It has herbs in it to fight the pain."

Even the slight movement of leaning forward sent my head reeling, and the pain doubled in force. I moaned.

"I know it hurts something awful, but your arm has started to heal. The cats have seen to that. They've been doctoring you."

"Right," I said as Frall lowered me back down. "As they dined on my flesh?"

Half the water had gone down my chest. I shivered under the fur and clutched it to me with a grip that not even Frall could have untangled.

"Clista, what are you talking about?" Frall's eyes examined me as if puzzled. I couldn't stand to look at him. I shut my eyes so I wouldn't see his lies.

Frall shook his head. I felt it even when I couldn't see the movement. He reached up and touched my forehead again. "Yte and Merot didn't cause that wound, Clista. You remember that, don't you?"

I opened my eyes to gaze into Frall's. He was studying me, looking like my words had confused him. "Frall, I remember everything," I told him sadly. "How you let the cats drain me of blood over and over. That hurt unbelievably. Why you would do that . . . Why would you let them…. I don't understand why you did that. I thought you loved me."

Frall grabbed me up and hugged me to him. "Oh, Clista. How can you ever doubt me? I love you with a fierceness that takes my breath away. You can't possibly understand how worried I was. I'd never hurt you. Don't you know that?"

He was hurting me now with his wild embrace, but I didn't say anything. The dizziness had come over me again, and I was concentrating on not throwing up.

"Frall, let me go," I finally said, and he released me as quickly as if I'd hit him. It was lucky he did. I erupted. Of course, there was almost nothing inside me, but I vomited up all the herbs he'd forced down me.

"Darn," Frall said. "I'll have to make more."

I guess our conversation would have continued then, but I was too weak to argue. My eyes closed, and I fell back into semi-consciousness.

Frall says I slept another day. I don't remember. The cats licked at my bloody arm. I know that. They woke me several times. I moaned and cried, but still, they kept at it. Frall never chased them away. Once more, he betrayed me.

The next time Frall doctored me with the herb drink. I kept it down, and the pain did actually subside a bit. I began to think that I might live after all.

Merot spent a great deal of time at my side. His tongue was frequently licking the wounds on my arm or my face. I enjoyed neither. Nor did I savor the ministrations of Frall. I felt cut off from all my friends and loved ones. That's when despair entered my heart, and I grew despondent.

Frall tried to get me to eat, but I didn't want anything. I turned my head away. Merot nipped me.

"Ouch," I yelled, and Frall, rather than giving me sympathy, rammed a piece of meat into my mouth. He'd been chewing it, and he delivered it from his mouth into mine. I was disgusted and almost spat it out, but he held my lips closed and forced me to swallow.

When he released me, I flared at him. "That was revolting. What's wrong with you?"

"You must eat, yet you keep refusing. Merot says you've lost the will to live. Why, Clista? Why don't you want to eat? You have to regain your strength. I won't allow you to slip away."

"Why do you care?" I hissed, furious with his betrayal.

Merot growled at me. Yte's paw slapped my face. I shut my eyes, but that didn't stop me from seeing Frall's expression. He'd deflated as if I'd pulled his plug. His eyes held tears.

I turned my face away from all of them, but Frall wouldn't allow that either. He kept on chewing the meat and spitting it into my mouth, then holding my lips closed until I swallowed. It was not a pleasant way to dine. My eyes glared, and my hands grew fists.

The rest of that day, Frall attempted to talk with me, but I turned my head away. I wouldn't listen, and hate poured from my eyes.

That evening, Frall walked me to the bushes, and I squatted to pass my water, but I was so weak he had to hold me. It is difficult to need someone when you're busy hating them.

The night passed in restless, unpleasant sleep. Nightmares haunted me. When morning came, I was relieved to see the light. I also saw then what was left of my arm. It was a red-striped horror of puckering and oozing wounds. I cried at the sight of it.

But when Merot plopped down beside me to torture me with his tongue of pain, my tears had been emptied. They still burned in the back of my eyes, but no water poured down my face. I hated too strongly to produce more tears.

My hunger was back that day. I ate what I was given. Merot purred. Yte licked my face. Frall again gave me the herb water and sponge-bathed my body. I wished I could fight them all, but I couldn't. I turned my head to the other side and pretended that this wasn't happening to me.

That night, Frall lay by my side. His arm crept over me. I stiffened from his presence. I didn't want him beside him. Yet, I said nothing. I closed my eyes. Eventually, I slept.

Chapter Ten

Later, Frall untied my arm and built a smaller platform for me to rest it on. Then he urged me to sit up. For a while, I did, but it made me dizzy. I didn't want any breakfast, but I drank the herb drink. I knew Frall would force me to do so if I argued.

Yte padded over and lay down beside me. She put her head in my lap, and I couldn't help it when my fingers moved automatically to her chin. I stroked her and listened to her purr. Perhaps that was the best medicine of all. I felt stronger by the time I took my walk to water the ground.

"Here, either eat this, or I'll chew it for you and force it down your throat," Frall said roughly as he thrust a piece of meat into my hand. I glared at him, but he wasn't looking at me. His eyes were focused on the fire, yet I saw a tear streak down his cheek.

My stomach growled. I brought the meat to my mouth and bit off a piece. We had no salt and no spices, yet it tasted delicious. Hungrily, I devoured it. Frall brought me a cup of water then, and I drank that greedily.

"All right. We're going to talk about this now," Frall said.

"I'm tired. I want to rest," I responded dejectedly. My anger felt as sore and painful as my arm. I worried that I might vomit again.

Frall grabbed me, his fingers pressing in against the flesh of my uninjured arm. "No, I said we're going to talk about it, and we are by the stars, Clista. I can't take your coldness anymore. What have I done to you that you hate me so?"

My tears started. How could Frall be so false and so cruel?

"You know why I hate you. You let them . . ." I sobbed, hiding my face from him, wishing he'd let me die.

Frall forced away the hand that hid me. "You mean when the cats licked your arm? I told you why they did that. A jasmire's claws are poisonous. You would have died, Clista. They saved your life. They treated you as one of their own."

I froze, and I listened. A small piece of anger plopped to the ground. It sizzled there, lessening a fraction of the acid burn inside my heart.

"They did what?" I asked again, trying for the first time to understand. Frall sighed and explained. More pieces of rage chipped away. I lost count of how many fell to lie amid the granules of rock and soil. Their loss lightened my heart. I sat up straighter. I turned to peer inside Frall's soul. My eyes dared to hope.

"They tortured me," I told him. "You let them."

"No. Clista, no. They saved your life."

My mind was trying to understand. So many scenes rushed through me —Yte on my chest, Merot licking me. Had they really been trying to save me? Had they taken my blood only to heal me?

"They had to do it?" I whispered, looking up at Frall, searching his eyes, trying to reach into his mind.

"You doubted that? Oh, Clista. How could you believe. . .?" Frall threw his arms around me and pushed my head onto his chest. "Stars,

Clista. We've all been sick with worry. Didn't you realize we were all working to save your life?"

I sobbed then, letting go of my fears and releasing the rest of my anger. Then I scooted closer to him. My arm burned when it felt the movement, but I ignored it. I was used to pain. What I needed was Frall's arms around me. I needed what we'd had before.

Frall patted my hair, brushing it back. It was dirty and full of tangles. I was embarrassed. I'd forgotten. I must smell. I must look a sight. I pulled away.

"I need a bath," I said. "I'm dirty, and ..."

"Stop it, Clista. You've been sick. Now you're getting well. None of that matters. I'll take you down to the brook, and I'll wash your hair when it warms up in the afternoon, but I don't want a word about that now. I just want to hold you, to know that you're okay and that you don't hate me anymore."

Hate? I felt a rush of love so strong I almost didn't quite know what it was. For a moment, I thought it meant that I was going to be sick again. But then I recognized it. I was simply flooded with joy.

"Oh, Frall. I love you so much," I cried out, and my tears wet the dark hairs on his tanned chest.

He took my chin in his hand and held me still. "Do you know how much it means to me to hear you say that? Do you really understand?"

I couldn't answer because Frall's mouth had covered mine, and he was kissing me. It was a gentle kiss, not one of passion, but it went deep into me. It was a kiss that a man who truly cared gave to the woman he loved, whether her hair was matted, her breath bad, or her body saturated in sweat from days of lying ill. It was a kiss that told me more about Frall than I'd ever realized.

I kept that knowledge inside me. I didn't want to talk about it, not yet. It was a jewel to hold, something to be pondered and savored.

"Why didn't the necklace work?" I asked a few minutes later when I was able to think clearly again. The query sprang out of me, not related to our words, yet I'd been wondering that question throughout my illness. What had happened? Why hadn't the jasmire recognized the *Old One's* tongue? Why hadn't it feared my power?

"I don't know, Clista. I don't understand that either. And Merot and Yte have no explanation. They spoke into my mind. We are all puzzled.

"They'd never seen the strange jasmire. He'd never been in their territory before. Merot said he smelled strange. I don't understand exactly what that meant. Our communication wasn't really clear.

"The other mystery is the *Old Ones* themselves," Frall told me. "I was sure they'd pull us back if one of us got hurt. I didn't think they'd ever risk you like this, but they must not have been watching. I can't believe they wouldn't have recalled us if they saw what occurred."

Merot came galloping over. His mouth was grinning. His purr was loud. He sat down beside me and rubbed his muzzle on my good arm.

I reached out to touch his head, but he ducked away and walked over to my arm. He sniffed and licked it once or twice.

"Eeooww," he said, and he returned to allow my hand to touch his head. Then he collapsed in a heap at my side and pushed at me. Obediently, I began to rub underneath his chin. His purr grew even louder.

Yte followed soon after. She was dragging something. She'd made a kill. Frall stood up and set to work, stripping off the prey's fur and preparing it for cooking.

"Buff?" Yte questioned me.

"I'm fine," I told her. "Frall told me what you both did. Thank you."

Yte brushed her head against mine. Her purr, so different than Merot's, sounded raspy in my ear. With Merot on the other side, it was like hearing stereo with faulty speakers.

I laughed and wished I had two good hands, but I could only pet one cat at a time. Yte licked my face, letting me know she understood.

Frall cooked the meat Yte had brought, and we ate. I drank some more of the herb, and then later, Frall carried me down to the water as he'd promised. There was nothing to shampoo my hair with, but Frall washed it the best he could. I scrubbed the rest of me with sand and water and sighed at the pleasure of doing so.

The cats didn't want me to get my wounds wet, but Frall had untied the limb from my arm, and he let me dip it into the water. The coldness took away the pain. I let my wounds soak until I was so chilled my body was a wall of goosebumps. Then Frall carried me back to the fire and helped dry me with his shirt.

That evening, I felt so much better. I even laughed. It felt good to laugh. Laughter is the vacuum that empties you of all despair and replaces the emptiness with happiness.

Of course, maybe it wasn't just the laughter. Frall's arm around me and his occasional kisses helped my mood a great deal, too. I was full of love. It overflowed, making me feel light and free . . . and blissful.

And that night, when we lay down in our usual spot, everything seemed different.

The night was dark and slightly damp with cold. No moons or stars painted the sky. The tree canopy was absolutely still. Not even the wind played the leaves, and the monkey birds had been ominously quiet since the arrival of the strange jasmire. Perhaps they had journeyed away and had not yet returned.

"Thank you for taking care of me when I was so sick," I whispered to Frall.

"Of course, darling," Frall told me, kissing my eyebrow because he couldn't see to find my forehead.

"You even cleaned me up."

"Yes, but it made you pretty mad, didn't it?"

"I'm sorry. I was stupid."

"You were sick, Clista."

His hand moved about, searching for my hair. He was so sweet. He'd never once taken advantage. I sighed and moved closer.

"Did you look at me when you. . ."

He chuckled. "My thoughts were pure, Clista. I was too scared about losing you to think about anything else. Now, if I were to do that today, knowing that you were okay…"

I slugged him playfully. "Don't even think about it. Forget you saw anything, please?"

"I have to forget everything?" he asked, pulling me closer.

"Owwww," I cried out from the sudden burst of pain.

"Stars, I'm sorry, Clista. Want me to put the branch back on?"

My arm was suddenly a throbbing wall of pain. It had awakened sharply.

"Any more of that herb?" I whimpered.

Merot got up with a yawn and moseyed over. He sniffed my arm and settled down to give it another cleaning. I gritted my teeth and tried not to cry.

When Frall returned, I sat up carefully so Merot wouldn't decide to punish me with a quick, warning nip. Then I sipped the drink as the cat continued to torment my poor, aching arm.

"Merot is really irked with me," I told Frall. "He says I shouldn't have immersed my arm in water because that washed away all his saliva."

"Yes, he's given me the same lecture," Frall said with what I was pretty sure was a grin.

"But the icy water felt so good," I argued. "That was the first time since the jasmire attacked me that my arm didn't ache."

Frall took my hand in his. "If there was anything I could have done to take away that pain, I would have."

I smiled, even though I knew he couldn't see. The fire gave some light, but Frall's body blocked most of it. It probably wasn't a very good smile, anyway. Merot was being rather rough, or at least it seemed that way. His tongue felt like it was ripping apart any healing my skin had started.

"Ouch, Merot. That hurts, you know," I told him.

The cat didn't bother to respond, not even to send me a message. If anything, he only licked more fiercely. I hoped the herb would start working soon. The pain was so great I was about to jerk away, regardless of the consequences.

"Yeeow," Merot paused to say.

I took it as a warning and gripped Frall's hand more tightly.

"Darn it, Merot. Ease off," Frall told him.

Merot growled, but I think his tongue became the slightest bit gentler — either that or the herb had finally started to work.

I lay with my head against Frall's chest and grimaced, but yet, I was happy. He hadn't betrayed me, and I loved him, and I was still alive, and I would heal — eventually.

And, after all, I had no reason to complain. I thought about what could have been. That jasmire had been so ferocious. If Merot hadn't come when he had, I'd be dead.

"Merot, I'm sorry I keep complaining. Thank you for saving my life."

The cat stopped licking and looked up at me. It walked over to my head, brushed against me, and before I could get my hand untangled to scratch him under the chin, he returned to my arm. He thumped back down and began once more, coating my wound with cat saliva.

I fell asleep that night, cozy with happiness. My arm ached, but it was a quieter pain than before. Frall's arms were around me. His breathing was like the purr of the cats. That, and the fire and the cats at my feet, sent me into a pleasant sleep.

Chapter Eleven

"What in the stars did you do you do to your arm, child?"

Thoric's voice woke me. I opened my eyes to discover that I was in a room I'd never seen.

"You brought us back?"

"That should be obvious. Frall is sitting over there, sulking."

I lay on top of a white metal or plastic table. A large bird-headed instrument was scanning my arm. After I decided that it wasn't hurting me, I glanced over at Frall. Thoric said he was sulking? No way. Frall was furious. His gaze could have cooked a steak faster than a fire could.

"What's wrong?" I asked him.

"Nothing," he snapped, his face darkening as if I'd said something horrid.

"I left you two in a perfectly safe environment, and now I find little Clista's arm half eaten. That's nothing? You were supposed to take care of her, Frall. She is your responsibility. You knew that."

"Stop it. How dare you say that," I snapped at Thoric, glaring at him.

“Lie still, child. The machine is not finished assessing the damage.”

“Then stop blaming Frall. If anything, it’s your fault. You dropped us there with no tools or weapons to protect ourselves. It’s a miracle we’re not dead.”

“For a pocket-sized queen, you certainly have a large temper.”

“I don’t like your insults,” I told him.

“Insults? Do not be silly.”

Frall bolted up and came over to stand beside me. He kissed me. “At least you’ll get some medical treatment for that arm now, Clista. No more cat saliva, right?”

“That is the most barbaric thing I have heard of. Cat saliva? You allowed cats to lick it? Is that what you’re telling me? It is certainly a good thing I looked in on you two. Clista could have died from such treatments. Have you never heard of soap and water? Has your primitive existence on Rwawan . . .”

“Rwawan is not primitive,” I interrupted. “And we didn’t have any soap. You sent us there with absolutely nothing, remember?”

“Rwawan heals wounds with cat saliva, and you say it is not primitive?” Thoric asked, raising one eyebrow in a dance of disbelief.

“The jasmire has poisonous claws and fangs. What would you have me do? Let Clista die?” Frall yelled.

“Temper, boy. I will not allow your disrespect. You still are nothing but a cub. Your job was to take care of Clista. Why did you leave her? That was your mistake.”

‘That wasn’t Frall’s job!” I cried out, bolting up in spite of the machine.

Thoric glared, pushed Frall out of the way, and started up the machine again. "Do I need to put you under?" he demanded, eyeing me with a warning look.

"You can't. I'm part Old One, remember?" I said, holding his gaze firmly.

Thoric laughed. "Well, I suppose it is possible that she was too much for you, Frall. Was that the problem? Although a grandson of mine who can't handle. . ."

"I went hunting," Frall said, his voice icy with anger. "A jasmire attacked her while I was gone."

"You left her alone in the forest? Then you are a bigger fool than I thought."

"Stop it!" I cried, but I couldn't move. Thoric was holding me still.

"See how easy it is, Frall. You do not need to be rough with her, just strong. Use those muscles of yours, or are you afraid of her?"

"Send us back to the forest. I can't take this," I said.

"Get your hands off of her," Frall ordered, barreling into Thoric, but Thoric didn't budge. He reached into his pocket and pulled out an instrument that he pointed at Frall. "Maybe I need to send you on a couple of separate voyages. Nelona said that taught Clista a little respect."

"No, please! Please don't separate us," I begged.

"Ah, the queen has grown some wisdom. What about you, Frall?" Thoric said, his voice sharp as a Jasmire's tooth.

Frall glanced at me. I'm sure he read the pleading in my eyes. He sighed. Then he reached out and touched my hand and nodded. "All right, Thoric. You win. It was my fault. I should never have left Clista alone. But let's get beyond that. Can you heal her arm?"

"Of course," Thoric said, turning the machine back on. "If I can get her to stay still long enough to get an adequate reading, the healing will be done in a matter of minutes."

"Let him fix your arm, Clista," Frall said, meeting my eyes. "Then we can argue whatever needs arguing, although it won't do any good. He won't listen. I can tell that."

I nodded and lay still, my eyes watching the machine as it started its rotation around my arm and across my body. Thoric touched a button on the side of the plastic hospital bed, and the thing oozed out a blue spider's web of traction.

I screamed.

"Like it or not, my queen, you will be fixed back into prime condition. *I* will not permit such injuries."

Thoric then turned to stare at his grandson. "Come, while she is healing, you and I have some serious talking to do."

"No, please, don't leave me like this," I cried, frightened about being tied down while a huge machine hummed its way around my body.

"Of course, I won't leave you, Clista. Didn't my grandfather just tell me never to do so?" Frall said as he returned to my side. He leaned down to touch my face with his lips. "My grandfather's right about that, Clista. I won't ever leave you alone again."

"Well, it is good that you have learned one lesson, but you have many more to learn, young cub. Many more," Thoric said as he walked over to the room's exit, waved his hand, and the door slid open. In a moment, he was gone from our sight, and the door slid closed.

"What about Merot and Yte? We never even got to say goodbye to them," I complained.

Frall laughed. "Clista, you amaze me. We've been dumped into a whole different location. I don't even know if we're in Thoric's house. Do you?"

I shook my head. I hadn't thought of that. I looked around the room, searching for a window or for something to tell me where we were.

"And even if we are back at his house, he may spin us off into someplace totally different. We can't prepare for it, Clista. We can't get our feet steady. Everything keeps shifting. And all you can think about is whether we'll see the jasmires again."

"But they were our friends. I never got to say goodbye."

The machine beeped. I wondered what that meant. I guess nothing since it started a different path, veering off to examine my feet and legs.

"You don't think Thoric is going to send us back to the forest, do you?" I asked sadly.

Frall bent down and kissed my brow. "No. It's my guess his experiment went wrong, and he doesn't want to admit it. I don't believe he knew the jasmire had poisonous fangs. I think he miscalculated. But what he'll do next, I haven't the slightest idea.

"I wonder if Nelona and Goor are still here," I said.

I sighed. The machine had just beeped again and changed direction. This time, it was circumnavigating my middle section.

"Why does this thing have to examine my whole body? I thought it was only supposed to fix my arm."

Frall thought about that, his eyes watching it. "Maybe it's checking that the poison didn't do any damage to the rest of you. That's a good thing, Clista. The one thing we can depend on is that their technology is superior to ours."

"That's not the only thing," I told him.

"What? What else about them can we depend on?" Frall asked with a resentful laugh.

"We can depend on each other. You do trust me, don't you?"

Frall grinned broadly. "Don't be an idiot. Of course, I trust you. You defended me even against an Old One," he said, laughing warmly.

"No, I mean it seriously. We do trust each other now. That's something dependable."

Frall's hand caressed my face.

"Yeah, but your trust is often a little shaky," he chuckled, remembering my doubts about him.

I nodded. "You're right. I've distrusted you too often. I'm sorry. I guess that's what I needed to learn. But I trust you now, Frall. I'll try to have more faith in the future."

"Is that a promise?" he asked, chuckling again.

The machine beeped and moved upward. Frall edged over slightly so it wouldn't ram him. The tip of the machine was finally concentrating on my arm, the uninjured one. As it scanned me, it purred like a sleepy cat.

"I miss them."

"Who, your uncle and Franca?" Frall asked, his eyes moving back to mine.

"No, I mean. Yes, I miss them, too, but I was thinking about Merot and . . ."

"So you and he were soft on each other," Frall kidded me.

The machine beeped again and rose up to focus on my head and neck.

"I don't like this, Frall," I said.

"It'll be finished in a minute. Then it's going to fix your arm, and you won't be in pain anymore. Then you can hold me again," Frall laughed.

We were both silent for a minute, watching the machine. Its purring had gotten louder. I closed my eyes. "Tell me when it's done, please."

Frall had withdrawn his hand from my face. He wasn't touching me at all for fear of interfering with the machine. I missed his hand and the contact I'd had with him a moment before. I let out a whimper, not meaning to.

Frall sensed my fear and started talking. "I'm going to teach you how to use a bow and arrow, Clista. We should have made one before. We would have if the cats hadn't joined us.

"I can shoot pretty good with one. I know I can teach you. I'll have to stand behind you and, put my arms around you and mold your body into position. That will be the fun part. Then, when you learn how to shoot, I'll be able to take you hunting, and I won't worry so much about your getting hurt. You'll like that part, going on a hunt. I know you want to be able to fend for yourself. I'll teach you how to do that, Clista.

"I know how to construct a sling, too. That's something else we can use. It won't bring down a large animal, but it will be okay for the small ones. With the bow and arrow, you can even kill the large ones, even a jasmire, if you have to, darling. Okay?"

The machine beeped and moved down my right side.

"Thank you," I told Frall, opening my eyes and meeting his. "Thank you for being here with me. I don't know about the bow and arrow. I'm not much of a hunter, Frall, but I'd love it if you'd show me how to shoot one."

The outer door slid open. Thoric came striding in. "Is it done yet?" he asked.

"Well, it's not humming anymore. Why was the machine examining her whole body?" Frall asked. "I thought it just needed to repair her arm."

"Silly cub. How could something repair a section without knowing the whole?" Thoric sneered at him. "Let us see what it found."

The Old One walked over to a separate machine, one that was part of the wall. His hand touched the surface, and the wall to the left lit up. "Counts are right on schedule. No infection," he said. He turned around to look at me. "If there really was any poison in her body, it has all mysteriously evaporated. I rather doubt she ever had any."

"I told you; the cats' saliva neutralized it," Frall told him, gritting his teeth.

Thoric didn't answer. He was busy talking to the computer. He spoke so quickly that I couldn't understand what he said. From Frall's expression, I could tell he didn't understand the words, either.

Then the Old One turned around and looked at me. "You're also not pregnant, little queen."

I gasped and sat up. Or I would have if I hadn't been spider-webbed down to the bed.

"Of course, she's not. Don't insult either of us," Frall said, his face pale as white ice.

Thoric laughed. "All right, here goes." He spoke again into the wall. The scanning machine retreated, and a second robotic arm raised itself out of invisibility.

"Is this going to hurt?" I asked.

Thoric studied me for a minute. "Yes. I imagine it will. I'll put you under."

I had no time to think about that. Thoric raised his hand, pushed a button on a small handheld device, and I went limp.

When I woke up, Frall was carrying me. "What?" I asked, although that really didn't make any sense.

The sun was dreadfully hot. I opened my eyes. We'd been transported again. This time, we were on the beach. The ocean waves were slapping the sand. There were sea birds flying overhead, and the smell of salt and fish flavored the air. Everything was so bright my eyes watered.

"Put me down, Frall. It's too hot to carry me," I said in a low voice, one almost incoherent from its drowsy slur.

Frall didn't stop. He continued padding through the soft sand and climbed up the side of a slight incline. Sweat dripped off his brow. His shirt stuck to his chest. His breath came in short puffs.

"I'm awake, Frall. Put me down."

"Hold on a minute, Clista. I'm almost at the top."

The top of what? Where were we? I started to ask, but Frall slipped and almost fell.

"Darn, I'm sorry, Clista. Is your arm okay?"

I was clutching Frall's neck with both hands. I'd completely forgotten about my wounds.

"I'm fine," I told him, amazed that it didn't hurt to move my left arm.

Frall laughed and set me down. "All right. Walk then, my girl. Let's see how steady you are."

Steady, I was not. Whatever Thoric had done to me, the after-effects were dizzying. I clutched at Frall and waited a moment for the circling stars to pause.

"That's what I thought," he laughed. "Thoric said you'd be wobbly for a few hours."

"Just give me a moment," I said. "I'll be okay."

"Look, it's not so bad now, Clista. We're up over the knoll. I figure we'll head for the palm trees. That should give us some shade."

I followed his finger, saw the trees, and nodded. Then I took a step. Frall reached from behind and swung me back up into his arms. "Forget it, Clista. It'll be faster this way."

"But I'm too heavy," I protested.

I think Frall was saving his breath. Normally, he would have issued some snide remark about that, but he only kept walking, not stopping until we'd reached the shade. Then he set me down and dropped himself beside me.

"Stars, what I'd give for some water," he said.

"Do you think those are coconut palms?" I asked, looking up into the tall trees.

"How would I know? This is the closest I've ever come to a palm tree."

"Yeah, me, too," I told him. "I guess if they were the kind that had coconuts, we'd see some lying around the bottom. Right?

"I wonder why it's so quiet?" I added, worrying about ferocious animals like the jasmire that had ripped up my arm.

We were on the edge of a jungle. The palm trees were the opening to it, inviting us inside its depths. What kind of animals did jungles have? I was pretty sure I remembered there were lots of snakes, the kind that wrapped themselves around you, the kind that poisoned people and then ate them, the kind that dropped on you when you least expected it. I shuddered and grabbed Frall's hand.

"We probably scared everything away," Frall said. "I bet any animals that live here have never seen people before."

I nodded my head, more from fatigue and dizziness than from agreement. I hoped we'd scared them away permanently. "Do you still have your knife?" I asked.

Frall smiled. "Of course. You know it always stays with me — ever since the forest, that first forest, I mean."

I sighed. "I wish Thoric had sent us back to Merot and Yte."

"Yeah, me, too." Frall's eyes were searching the area. He didn't look any happier about it than I was.

"Did Thoric say anything before he sent us off?" I asked.

"Yeah, plenty. He lectured me really well," Frall laughed. I could tell he was still mad about it. His eyes darkened with the remnants of his irritation.

"What did he lecture you about?"

"You know, Clista. I'm never supposed to let you out of my sight. That kind of stuff."

"The Old Ones need to back off. They're going to make you hate me," I groaned.

Frall took that as a reason to throw his arm around my shoulder. "Clista, you know that's not true. No matter what they say or do, there's no way they'd change the way I feel about you."

I smiled at him and darted a kiss. I hoped he'd keep saying that. The Old Ones sure interfered a lot. I leaned back against the tree and yawned. "Would you mind if I took a little nap, Frall? I don't know what's wrong with me, but I'm really sleepy."

Frall grinned. "Thoric said you'd probably sleep for most of the day. Sure, go ahead. I'll be right here. I will never leave you unattended again, remember?" he said, chuckling.

I felt him shift me around so my head was in his lap. It was so comfortable. I yawned again and then fell asleep.

It was evening when I woke up. Frall was still watching me. In fact, I think he'd been tickling me with a piece of palm frond.

I sat up.

"Oh, my legs," he groaned, and then he laughed. "Turn around, would you. I had to wake you up. I'm sorry, but I desperately need to use good, old Mother Nature."

The air had cooled. A breeze was blowing. It felt wonderful. I stood up and turned away. The dizziness was completely gone.

"Oh, I feel so much better," I called out.

"Good," he said as he returned. "Let's see if your legs will carry you enough to search for some water. I'm thirsty enough to drink the ocean right now."

"It's back that way," I said, pointing the way we'd come.

"Funny, ha! ha!" Frall said, pulling me forward.

We walked for about a mile, following a trail some animal must have made. Frall said it was an old trail. He told me that nothing had traveled it in a long while.

"How do you know that?" I asked, turning to look up into his face as if he had such things written across his forehead, chin, and cheeks.

"The tracks are faded. There's no fresh poop, and the bushes are starting to cover it up again."

"Wow! You do know about stuff like that. I'm impressed," I said, my mouth dropping with surprise.

"Courtesy of Yte," he said, grinning. "She also called me "cub," I'm afraid, and told me frequently that I walked like a human. That was supposed to be an insult, of course."

I laughed. Then I stopped. "Listen. What's that noise?"

Frall looked at me strangely. "Are you okay? I don't hear anything except for the… the roar of … water! Come on, Clista. I think it's a waterfall."

Frall must have been really thirsty. He galloped forward, pulling me along. I could smell the water as we approached the fall. It was in the air, in the moisture all around us.

"Oh, my stars!" Frall exclaimed, slamming on the brakes as he parted some bushes and stared beyond them.

"What?" I ask, coming up from behind.

Frall pushed me in front. "Stars!" I echoed him.

The water dropped from about thirty feet down from the side of a huge cliff. The run off jetted outward and then ricocheted off several different overhangs. Then it fell, thundering worse than a military jet, in a mighty cascade that took it straight down into a sky-blue pool.

"So beautiful!" I said, breathless from the wonder of it.

"Yeah, and it's running water, so we can drink it. Come on, Clista. I'm dying of thirst."

We made our way down the path that headed right to the little lake. I know we should have tested the water, but how could we? Frall had his head in it before I could even caution him. And he was slurping up handfuls of it.

He stopped and looked up at me. "Aren't you thirsty? It's delicious. Come on, taste it."

I was parched from our run. I bent down and drank. Frall was right. Nothing had ever tasted so good before.

The waterfall was sending moisture all around us. Frall's hair glistened. My face dripped from the spray.

"I think we've found paradise," Frall said. Then he stripped off his pants. "I'm going for a swim, Clista. Want to join me?"

I wasn't that adventurous. Not yet, anyway. I shook my head, looked the other way, and listened as he dived in. Of course, he splashed me. I scooted back.

When Frall came up, his face had the smile of the boy I'd known back at school. I sighed, happy to be there watching him, happy to see that smile again. "I love you," I whispered, but of course, he couldn't hear me. He was heading across the pond, swimming closer to the drop point of the waterfall.

I sat down, took off my shoes, and dipped my toes. The water was cold, but only pleasantly so. I sighed from the beauty and the contentment of being in the peacefulness of this moment. The water mist carried in the air seemed to bathe me with a nectar of forgetfulness. All that had gone before seemed far away. It was only the here and now that mattered.

Green fronds surrounded the rocky out hangings of the pond. Flowers bloomed —violets, orange nasturtiums, and something I'd never seen before, with dainty petals and a deep, dark crimson color. The scent of the damp soil tickled me. It smelled of rain and rich, pungent vegetation. I breathed it in and let out a contented sigh.

I scooted closer to the water and let my feet drop lower. As I did, I carelessly used both arms to scoot me forward, not thinking about my wounded one. Afterward, I winced automatically, expecting the pain to hit, but it didn't. My arm felt normal.

I investigated. I'm not vain, but the sight of the wounds on my skin had formerly turned my stomach. I'd tried not to think about the way my skin would finally end up when it healed; sure, I'd have horrific scars for the rest of my life. What had the machine done to my arm? Did it look worse?

With shaking hands, I tried to roll up my sleeve to see, but the dress I was wearing didn't permit that. Not unless I tore the material at the cuff. Not knowing how long Frall and I would be banned from this particular location, I certainly didn't want to destroy my one and only outfit.

I took another look to see where Frall was. He was still heading for the place where the falls hit the pool. I slipped the dress up over my head and set it down on the bank. Then I forced myself to look at my arm.

It felt a bit stiff, and the skin was overly tight — just about like it would have been if I'd gotten a bad sunburn. But there was no sunburn and no scar. My skin was baby-new, perfect. Nothing remained of my scratched and torn flesh, no stitch marks, no gashes, or even redness. I gasped and studied it, turning my arm this way and that, blinking and looking again in disbelief. It was absolutely incredible. My arm looked just like it had before the jasmire's attack.

"Whoa! I believe I see an enchanted mermaid on the other side of this pond. I'm in love. No one but a mermaid could be that beautiful," Frall yelled as he trod water.

I blushed and reached for my dress.

But either Frall thought he had a chance of reaching my side before I got my dress back on, or else he was ready to quit swimming. For whatever reason, he began his return, swimming so fast it was as if someone was chasing him. I wanted to watch; his glide and stroke were stunning, but I turned and grabbed at my dress, sliding it down over my head. Then I discovered that it was inside-out. The sleeves were turned about. I had to take it off again to fix all that. Yet still, before Frall touched the bank where I was sitting, I was clothed again.

"Darn," Frall said, his smile drooping sadly. "Where did my beautiful, enchanted mermaid go? She was lovely!"

I laughed nervously, then said, "Stop it, Frall, please," feeling the heat in my face, knowing that I was turning several shades of red.

"But I thought you were ready to come into the water with me. Now you're dressed again. What happened?"

I shook my head. "I wasn't planning on swimming. I didn't take my dress off for that. I thought you were busy checking out the waterfall, and I just wanted to take a look at my arm. I tried to roll up the sleeve of my dress, but it was too tight. That's the only reason I took off my dress. I didn't mean for you …" I looked down, not finishing my sentence.

Frall didn't say anything. The water was slapping against the boulders. Its motion caused small waves. I looked up again to watch.

He was smiling, staring up at the clouds and the blue of the sky.

"Guess what, Frall," I told him, ready to forget the previous conversation. "My arm is completely healed!"

"Yeah, I know," he said casually, and he flipped onto his back so he could float.

I wish he hadn't turned away. I couldn't read his face. "How would you know that?" I demanded.

"Who do you think . . . Never mind, Clista. Maybe I'd better get out so we can hunt up some food. That's the next problem, isn't it? Then I suppose we'll have to figure out where to sleep, and . . ."

"You're not getting out of that water, Frall, until you finish your sentence. What do you mean you already knew about my arm? And how dare you say *never mind* about it. Tell me. What do you mean: Who do you think. . .?"

"I'm not supposed to get out of the water or what?" he asked, grinning and paddling about so he could study my face, although he was still leisurely floating on his back.

"I'll. . . I'll . . . take your clothes and hide them where you will never find them," I said, standing up and picking up his pile of garments.

Frall laughed so hard he sank slightly and swallowed water. He came up coughing and clutched at the rock overhang so he could catch his breath. The moment he did, he laughed again.

"I'm not kidding, Frall," I warned him, backing away. "Tell me what you meant."

Frall hung there on the side, still chuckling, watching my face. "Clista, your threat almost drowned me. Don't you care?" That sent him off into another spasm of laughs. Then, seeing my glares, he sobered. "Darling, do you really think I care if you toss away my clothes? Who will it bother more, you or me, when I rise up out of the water minus my cover?"

"I don't believe you," I said, backing away.

"Okay, Clista. I think it's time to come out now. You can either give me the clothes or not. It really doesn't matter, but we'll be spending a lot of time together. Do you want me walking around the island wearing my birthday suit?"

He wasn't wearing underwear? I backed up and watched as he sank down and then hoisted himself up from the boulders alongside the pond. When I saw that he was climbing out, just like he'd said, my taunt ended. I hastily dropped his pile of clothes and turned around.

Frall was still laughing. He padded over and picked up his clothes. I heard him shake the water off himself, just like the jasmires had whenever they got wet. Water flew everywhere, some of it on me.

I took several more steps, moving further away from him, but I was eager for him to finish dressing. I really wanted to know what he'd meant. There were other ways to get information. I'd just have to find the key.

Frall rustled the material of his pants, making noises like he was pulling them up. I heard the snap as he fastened them. I turned then and watched as he pulled on his shirt.

"Would you really have walked around without your clothes on?" I asked.

He grinned. "You'll never know now, will you, darling?"

"That's disgusting!" I said, looking away from his smirk.

"Yeah, but you still love me," Frall said as he walked closer.

I backed away again. "You still never told me. How did you know my arm was completely healed?"

Frall reached out and grabbed my shoulders. Then he pulled me in close. "Ah, my mermaid, remember that poor rag of a dress you were wearing in the forest? You still had it on when the machine was fixing your arm. Don't you remember that I was there? Of course, I saw the results when you were still unconscious."

I breathed a sigh of relief. "Then who changed my dress? Tell me that wasn't you."

"You think I would have dressed you in that thing?" Frall laughed again. "Dear Nelona came in and did the honors while Thoric and I went into the other room to chat about my behavior."

I sighed, so relieved that once more, my face blazed with heat. "I'm sorry, Frall. I guess I wasn't trusting you again," I said. I kissed him on the lips with more enthusiasm than I would have a moment before.

"That's not much of an apology, my mermaid. How about doing it right?"

I improved the kiss the second time. In fact, we were both enjoying the pleasure of it when a pudgwie started throwing some kind of nuts down at us.

"What in the stars?" Frall yelped when one hit him right on the forehead. We both looked up and saw the animal high up in the tree. The little guy was adorable — curly-tailed and big-eyed. Its silver stripe ran down its back and mirrored its tail. It made the little guy look like his fur was rippling, like water on a windy day.

"Oh, he's so cute!" I cried out.

"Yeah, wait until it hits you with a hard nut."

"You don't suppose they're edible, do you?"

"The pudgwie? Sure. Good thinking. I'll find a rock and see if I can…"

"No! Don't you dare?" I grabbed Frall's arm before he could pick up a rock. "I meant the nuts. What kind of nuts are they? Can we eat them?"

Frall sighed and shook his head at me. "We don't eat meals if we don't kill something, Clista, and I'm not going vegetarian."

He bent down and picked up a couple of the nuts to take a closer look. "I don't know what they are. I've never seen this kind of nut. They could be poisonous or delicious. I just wouldn't know."

"Okay, then let's keep walking. Maybe we'll stumble onto something we do recognize."

Frall gave a wistful look up, but the pudgwie was already gone. Frall shrugged, picked up a couple of rocks, stuffed them into his pocket, and took my hand.

"Any pain with the arm?" he asked.

"No, it feels great. Next time I see Thoric, I really have to thank him."

"Or curse him. We haven't seen what's on this island yet."

"How do we know it's an island?" I asked, wrinkling my nose in thought.

Frall shrugged. "Good point. I guess we don't know. We could be on the edge of a booming city, but somehow, I doubt it. I think Thoric wants us to be alone."

We continued walking for several minutes, not speaking, just looking around. Everything was super green, a green so emerald, it almost made your eyes hurt. There were blooming flowers everywhere, tiny splashes of color amid bundles of green, and some plants were growing grouped together like multi-colored bouquets. I oohed and ahhed at each new showing until Frall began to tease me by imitating my expressions of wonder.

"Knock it off," I complained, but I tried not to be so vocal. I guess it did sound silly, like people admiring fireworks: oooh, aaah, look at that, ohhhh!

The further into the jungle we walked, the thicker the patches of flowers and the more abundant the colors. We came to a plant with long spears of gray-green brilliant orange blooms that looked like tongues hanging down. That sent Frall into hysterics. He tore off one and handed it to me. "This flower you've got to love," he said. "Tongue waggers!"

I breathed in the flower's aroma, hoping for something like ginger or cinnamon. No such luck. It smelled like rotting garbage. I handed the thing back to Frall, wiped my hands on some nearby grass, and said, "No thanks."

Frall laughed. I don't think he believed my facial expression. He put the flower up to his nose and breathed in deeply, as if to show me how it should be done.

"Phew, that thing REALLY stinks," he exclaimed, throwing it down. "I'm sorry, darling. I never meant to give you anything so nasty."

"Well, next time, smell it first," I told him, laughing. "If it's the scent of dirty socks or perfume of sweat, choose something else."

He threw his arm over me, corralled me with the other, and kissed me. "That's making up for the stinky flower," he said.

It might have been more romantic if Frall's stomach hadn't rumbled while he was kissing me. I broke off, giggling.

"Darn," he said. "I've missed several meals, I'm afraid. Must be dinnertime, according to the sun."

I looked up. "There's only one sun. Did you notice that?" I commented.

"Which means we're not on the same planet as in our latest forest vacation," Frall said, nodding his head.

"Nor on Flagan Pren. It had at least one moon up in the sky at all times."

We continued on, still searching for something to eat. There didn't seem to be much available, at least not that we recognized as eatable.

"Old Grandpa Thoric wouldn't have shipped us off to a planet with no food. At least, I don't think so. There has to be something we're overlooking," Frall said.

"Well, there are fish in the ocean. I suppose there could be tide pools around somewhere."

Frall sighed and stopped. "That's a really good idea, Clista, but we're going in the opposite direction." He turned and looked back the way we'd come.

"You said you thought this was an island," I said. "Maybe if we keep going forward, we'll find the other side of it."

"Yeah, unless it's a big island," he laughed.

I sat down, tore off a leaf from one of the bushes, and began to fan myself with it. "I don't care what we do, Frall. I'm even ready to go back to the waterfall. I'm just so hot."

"Me, too. Tell you what. Let's keep going forward the rest of today, but if by nightfall we don't find anything we can eat, we'll turn around in the morning and head for the falls again. Deal?"

I nodded. "I wish we'd brought some water with us."

Frall nodded. "Well, the good thing about that is that at least we know there's fresh water on the island. If it's an island."

We were hungry, thirsty, and tired, but we smiled at each other. Frall took my hand and pulled me up. "Come on, mermaid," he said. "Let's see what there is to see."

A couple of hours later, we were still seeing the same thing, flowers, ferns, and huge leaves that looked like placemats. Sometimes, we caught glimpses of pudgwies, but none of them gave Frall a chance to throw his rocks. None of them threw anything at us, either. They just chattered and ran up and down the trees, running faster vertically than Frall, and I could run on the level ground.

We kept hiking until the light faded. Then we made a soft nest beneath one of the trees by stripping off some of its soft, flat leaves, putting some underneath us and others on top. Clutching each other, more for security than for warmth, we slipped into sleep.

In the night, we woke up several times. A jungle creature was calling out, its cry a half roar, half vibrating scream. We tightened our grips on each other, pulled the leaves over us again, and closed our eyes.

When we woke in the early dawn, it was raining. It began as a mist, but in minutes, it was a deluge that pelted us painfully. We held the leaves over our heads, scooted our backs up against a tree, and waited it out. The foliage of the tree we stood beneath helped somewhat, but we were still drenched and miserable. We did get our drinks of water. The heavy leaves held bowls full.

Frall's nose began to run. He used a leaf as a handkerchief. But as the sun finally heated the clouds away, I saw that he was getting sick. His eyes had reddened, his nose had turned bulbous, and in an hour or so, he'd begun to cough and sneeze.

"My poor Frall. We'd better get you someplace warm and dry."

"Yeah, a fire would be great," he said.

"I didn't tell Frall that despite all the trees and huge bushes we'd seen, we hadn't found any usable firewood, and even if we had, it would have been too wet to burn." With Frall's teeth chattering and his flushed and sickly color, I didn't dare mention that.

"Let's go back to the waterfall," I said, thinking that at least there, I could cool off his burning forehead.

He nodded. On our continued trek that morning, he gave no more than a couple of yesses and no other responses. By late morning, he was leaning on me, although I'm sure he wasn't aware of it. He was stumbling, too, and his forehead each time I checked felt hotter than the time before.

"Thoric," I called over and over in my mind. "Help us, please." I met with silence, but I persisted. "Thoric, help us. Frall's sick. Help us."

Frall collapsed about the time the sun was directly overhead. I couldn't lift him. I shifted him about until he looked more comfortable, and then I attempted to cool him off with leaves. Despite the rain that morning, the moisture had all dissipated, and although the humidity was high, the heat felt even more oppressive.

Frall began to thrash about and babble about school and my uncle. He was talking about a report he needed to finish, and he blamed Uncle Cristoff for taking it from him. He said Cristoff had stolen it from his locker. In his feverish condition, Frall vowed to search Smog's cabin, declaring that he would tear it up from bed to table until he found his unfinished paper.

"Frall, it's okay," I told him. "I've got your report. I'll turn it in for you."

At one point, Frall even hit me, saying I was to blame, but he called me Cristoff and thought I was trying to steal another assignment.

I removed Frall's shirt, trying to cool him down, but he fought me when I tried to pull it away. Then, seizing my hand in a grip that left me immobilized, he demanded I give him a kiss. Semi-lucid for that moment, he laughed evilly and claimed that I was a mermaid once more, attempting to steal his clothes. It was a brief period of coherence. Almost immediately, Frall cursed me and called me Cristoff again.

All that afternoon, Frall thrashed about. His fever burned. We had no water. My lips were parched, as were his. I attempted to ease his fever with a cool leaf. Frall bit it and began to chew.

"No, Frall," I yelled, and I tried to get it away from him, but he swallowed and wrestled the rest of the leaf away from me. When I tried to reclaim it, he slammed me so hard I fell. It stunned me, knocking the breath out of me. I lay where I'd landed for several minutes. My tears started then. I was at the edge. Thoric had given me more than I could deal with. I yelled at the *Old One*.

Then I shut my eyes and concentrated. I wished so strongly that Thoric were there with us. If I'd had any power at all, the *Old One* would have been staring into my face.

It didn't work. I calmed myself and recalled the necklace. In my eagerness to get at it, I tore the neck of my dress, but my hand found the metal and focused.

"Take us back to Thoric. Take us back to Thoric," I said over and over, chanting.

The rain started. I rushed to Frall's side and covered him with leaves. He was unconscious then, or at least not thrashing about or babbling. His eyes were closed, so I couldn't tell if he was sleeping or worse. I placed my head on his chest and listened. His heart was beating, but it seemed slower than usual. Was that my imagination?

I sat back on my heels, staring up at the pieces of sky that could be seen through the canopy of trees. Then, once more, this time, with Frall's limp hand held against the necklace and my two hands holding his in place, I concentrated even harder. The *Old One's* language came into my mind. I used words like those that had spilled out of me when the jasmires had come. I focused on my mental image of Thoric and on the appearance of his dwelling. I felt nothing.

Whatever strange animal had roared/cried before was suddenly gargling ominous noises again. I wasn't sure, but it sounded like the creature was getting closer.

Frall started thrashing again, his face dripping with sweat. His body felt like a potter's kiln.

"Can't. She won't. Will she? Cristoff… must get back the work… Clista, darling. Where are you? I love you. My homework. Where… my homework? Cristoff has it. He hates me. Clista, Clista . . ." he babbled on. Some of it was incomprehensible. Other words made perfect sense. I shut them out as best I could, gritting my teeth against my panic, determinedly focusing again on Thoric.

But my mind was distracted. I wanted to call out to Frall to reassure him that his work was fine and that I was beside him. No, I told myself. You cannot. You must focus.

I pictured the way Thoric's house had looked from the outside. I saw the high arches of the entrance, the manner in which they'd oozed into a welcoming door. I saw us going inside the inner chamber where the walls had closed us in. Then, I flew forward in time. I viewed the water fountain where Frall and I had played. What a great time we'd had until that day until Thoric had come and scared us half to death. Then we'd stood before him trembling, fearing we were in trouble. I pictured all of that, and I tried to take us there.

But nothing happened. When I opened my eyes, Frall and I were in the same place, lying in the leaves and dirt or some faraway jungle.

I shut my eyes and imagined the table on the second floor where Smog, Franca, my uncle, and Thoric and his wife had all sat down to dine together. I saw the throne room again, the place that had once frightened me so greatly. The two mighty chairs were gray with age and held black streaks in the marble from which they'd long ago been carved. I pictured the thrones mounted high on the platform, beckoning and enticing, wanting me to climb onto their stone-hard surface.

"Bring us back," I called to the twin thrones. "Let us sit once more inside your great expanse."

I opened my eyes. The jungle remained all around us, the sun casting ripples of light as the leaves of the trees shifted slightly in the faint breeze, a breeze that brought no relief from the humidity.

I visualized and thought and tried again and again. I called up and focused on the woman who claimed to be my grandmother. I sent out mental messages, but nothing worked, and in the back of my mind, where I really didn't need it, was the awareness that whatever it was that growled and hissed and thundered like the sky — it was approaching steadily.

My heart sped up. I gently placed Frall's hand down on his chest and turned in the direction of the noise. As I waited and watched, my eyes scoured, hoping to see something I could use as a weapon, but all the plants were spinelessly soft except for the trees, which offered no safety except height. Their branches were high up, obviously reaching for sunshine. For that reason, they offered no assistance for climbers — unless one were a spry, clawed pudgwie.

"Frall, can you hear me?" I cried, attempting to shake him awake while I stared into the bushes for a peak at whatever was hunting us.

Frall moaned, opened his eyes, and stared at me. "What's wrong?" he asked, slurring like a drunk, but otherwise, for the moment, lucid.

"Something's coming. I think it's dangerous. There's nothing to fight with. The branches are too high. I can't see any fallen ones. We can't climb. The magic is dead. I know you're sick. You can't help but. . ."

"Same animal . . . growling . . . in the night?" he asked, licking his lips, his yellowish, bloodshot eyes already beginning to glaze.

"I think so."

The roar sounded again closer, louder, and more horrifying than before. Goosebumps covered me. I shivered. The heat had not eased. Perspiration dripped. My brow was a stream, between my breasts, another river.

"Help me . . up," Frall ordered.

I bent down to grab his hand. He took hold of mine, gripped the tree with his other hand and pulled. He gave an effort to stand up that took every last big of his energy. But it didn't work. His legs refused to support him, and he began to sag.

"Clista . . . I can't."

"I know," I said, letting him slide back to the ground as gently as I could.

He was panting. His face had turned red, the exact color of the crimson flowers we'd passed. I placed my hand on his forehead. The fire inside him hadn't cooled. In fact, I feared it was higher.

"Could I use the knife?" I asked as delicately as I could. I hated the look that passed over his face. He studied me for a moment but said nothing. He just nodded tiredly. His hand reached down into his pocket, and he pulled out the knife. He took a moment to turn it about. Then he held it out to me.

"Don't… throw it," he told me. "Only. . . one chance. Odds . . . against you. Wait. Aim . . . with a thrust. The eye . . .is the brain . . . Stab hard."

"I will," I promised, and I tried not to think about Frall's words. Me killing? Me, the one attacking an animal?

My hand shook as I reached for the knife. Frall met my eyes. "No choice. . . be brave… Don't think."

I could see that he wanted to say more, but he couldn't. His eyes were watering. Another coughing bout started up. He held off a moment, sliding the handle into my hand and closing my fingers around it. But then he gave into what was obviously a painful session, with a cough both ragged and wet. When he finished, he sagged back against the tree. His eyes closed. I thought he might have lost consciousness again, but he forced his eyes back open and gazed at me.

"I love you," I told him.

Frall blinked his eyes and sighed. Then he mouthed, "Me, too."

The bushes rustled. I turned to see. A creature, like nothing I'd ever seen before, stared back at me. It had paused to stare, its eyes two violet holes in a mass of umber feathers. Its beak, bright, reddish orange, opened and closed as if it were speechless. Our surprise at the sight of each other was obviously mutual.

When it pushed its head the rest of the way through the green plant covering where it had chosen to hide, I saw its body. It walked on four legs and had a body that was jasmire-sized. But, although its furry pelt was tawny-colored and its long, black tail lashed about nervously, it was like no cat I'd ever seen. Its spotted rump had both feathers and wisps of fur.

"Go away," I yelled at it, and I pointed Frall's knife in the creature's direction.

It roared as if it knew I'd challenged it. With its cry, a huge, snake-like tongue crawled out and down, slithering forward among the decaying leaves and soft dirt of the jungle floor. I stared at its approach as it weaved its way closer and closer. I readied my knife. The tongue writhed a good ten feet, but then, apparently, it could go no further and stopped.

I was eager to dart forward and stab at it. One brownish snake-like tongue was not nearly as frightening as the animal connected to it, but Frall cautioned me.

"Wait," he said.

So I took a stance of readiness and clutched Frall's knife with more confidence and even a slight feeling of zeal. I wanted the attack over. I wanted to get it behind me.

"That a girl," Frall said and then dissolved into another round of coughing.

Meanwhile, the animal, whatever it was, gyrated its eyes about. I refused to look at them, but still, I saw, as I stared at the ground, that it took three steps forward.

"Get out of here," I yelled, stabbing at the air in warning.

The thing's beak opened, and the bird/cat roared again. Then its mighty tongue withdrew, coiling up like the string of a yoyo-toy.

I glanced back at Frall. His eyes had once again closed. How could he sleep with a monster roaring so loudly and obviously about to strike?

I called out to it. "Hey, if you want to attack me, do it. Or, if you want to help, go get some water for us. Friend or foe. You choose."

I heard a snicker from behind me. Frall was still conscious.

The ugly bird/cat shook its head. A couple of feathers dropped. The sun, a moment ago hiding behind the clouds, broke out and glinted down on one of them. The feather turned itself about, planted its tip down into the ground and began to grow. Almost immediately, a shoot headed upward. In moments, it filled out into a full-grown bush.

My jaw dropped. I could barely take my eyes off such a miracle, but I never lost my awareness of the ugly cat/bird. The bush began to jiggle and wiggle. I wasn't sure, but that there might soon be two monsters attacking us. I dreaded the thought.

The bush continued to shake itself. It rustled and stretched outward. Its fronds opened wide. Then, with a strange, plant-like puff, it burst into one of the odd tongue-like flowers that Frall had earlier given me to smell.

"Ugh," I cried as the smell of it wafted through the air.

Apparently, the monster didn't like the odor any better than I did. It bellowed horrendously, turned about, and galloped off into the bushes.

"Hey, isn't that your baby you're leaving behind?" I yelled after it. I hoped that wasn't true about the plant since it had decided to grow in our vicinity, but if it was, it was interesting to note that the cat/bird certainly appeared to have no motherly tendencies.

"Good job," Frall said, attempting to sit up straight again. It was only for a moment, though. His head slowly sagged, and his body folded in like an angel food cake I'd taken out of the food machine too quickly.

With the monster gone, I set the knife down next to Frall. He'd started shivering, so I retrieved his shirt and covered his upper body as best I could. I straightened him out, so he'd be a little more comfortable, and then I held a leaf over his head to prevent the rain from soaking him because it had started up again.

The leaf I was holding began to drip. I drank. It soothed my parched throat. For several minutes, I poured a steady stream into my mouth.

Then I turned the leaf to dribble moisture into Frall's mouth. He opened wide and took in the flow, swallowing greedily. At least one problem had been solved.

For the rest of that afternoon, the day dripped in coldness and misery; I snuggled up to Frall and warmed his body the best I could. His forehead and the rest of his body felt hot, but he was shivering. I didn't know which he needed, cooling down or warmth.

The cat/bird came back about the time the rain subsided. It watched us from the bushes. I hoped it was more curious than predatory. I yelled at it several times. Once, I even told it we were hungry and that it would be nice if it would deliver us a good, hot meal.

It blinked its violet eyes and roared. I don't think it liked my suggestion.

As the hours passed, and the creature didn't approach but continued to sit there in the bushes, blinking and staring, I suppose I got used to the danger of its presence and just accepted that it was there. I could do nothing about it, anyway. I had to remain at Frall's side. But as the afternoon turned into evening, I grew bored. What else could I do but talk to the strange cat/bird?

At various times, the animal let out one of its horrifying roars as if to dispute my words, but otherwise, it just watched and cocked its head this way and that as if listening.

"I wouldn't mind a cup of cocoa," I told it. "I'm not fond of the mint kind, but marshmallows would be nice. I really hate whipped cream. Don't you think that would go well with a big, steaming bowl of minestrone soup? No croutons. French bread, if you please. And absolutely no garlic on it. Garlic gives me diarrhea.

"Or, maybe just a plain, old hamburger would be better. I like mine with catsup, mustard, lettuce, and pickles. Cheese is good, too, I suppose, but I'd just as soon have it without. Frall likes his hamburgers with onions, but I hate onions.

My father likes — I mean, he liked — his with onion rings. That's kind of strange, huh? My mom never ate hamburgers at all. She preferred salads. It's not like she had to diet or anything. She was never even the slightest bit fat. She just really liked lettuce and vegetables. I don't mind them, but they're never as tasty as pizza, for instance.

"I'd ask you what you like to eat, but I might not like the answer, so I don't think I'll go there. How about your favorite color? Are you a blue kind of guy or a fan of green? I bet you like green better; green is like the jungle.

"Tell me something. Is this really an island? We still haven't found out. I wish we could have walked around with a backpack full of food. That would have been fun.

"A nice raincoat would be a pleasant thing to have. And a warm bed with blankets and a pillow. Darn, I'd even go to a cave right now. A warm cave with a campfire. Then we'd have some light and warmth and …"

The cat/bird stood up, shook itself, and then plunged through the bush, trotting toward me like we'd suddenly become friends. Frall's knife was in my hand the moment the creature moved.

"Don't come any closer," I warned it, waving the knife like it was a dangerous weapon the ugly monster should be afraid of.

But it didn't stop. Its eyes were focused on Frall, and it came to about three feet from us, stopped, and unrolled its tongue.

Frall had been silent for a long time. I thought he was still sleeping, but he suddenly sat up and shot one of the rocks he'd brought with him right at that advancing tongue. The thing coiled back up, and the cat/bird let out another of its terrible roars.

Instinctively, my hands flew to cover my ears, but the knife was in my right hand. That recalled me. I lurched at the monster, caught its skinny neck in my hands, and stabbed right in the eye. It bellowed in pain, shook its head, and threw me off. I landed hard, but that didn't stop me. I turned about and lunged at it again.

The monster's tongue was flowing down from its mouth, but the knife was still in its eye. I dived for the knife and slammed my full weight against its handle. Once more, the cat/bird cried out, but green blood was squirting from its eye socket. The animal was also shrieking with ear-splitting volume. I wanted to run from the noise, but I threw myself on the knife's handle and rammed it even deeper. The monster spasmed then. Its head bobbed, and its neck twisted. Then, its legs wobbled and buckled, and the huge cat/bird slowly collapsed.

Frall sat up weakly. He clapped once or twice and then fell back. "Cut throat," he urged in a raspy, hoarse whisper.

The limbs of the cat/bird still quivered. Its beak continued to open and close. Its eye leaked green goop all over the ground. I didn't want to go near the thing, but keeping a watchful eye on its tongue and its sharp, orange beak, I moved forward and took hold of the knife.

Pulling the weapon out was no easy task. I'd buried it deep into the creature's brain. I yanked and pulled. Finally, it released with a slurping suction that almost sent me head over heels.

The knife's removal caused an outpouring of the thick, green liquid. Like a geyser, the goop erupted upward and cascaded down. There was no way I could get close until that stopped. I stood and watched, trying not to gag, but the stench was appalling. The closest I could come to describing it was to say that it smelled like a mixture of fresh vomit and poop blended into a milkshake.

When the green geyser finally subsided, keeping an eye still on the bird's beak, just in case, I lifted the knife and brought it down again. At the point of impact, the bird jerked, and its beak swung around. I screamed and stepped back.

Frall called out, "Careful." He sounded so much like himself; I turned and looked. It was at that moment the bird's beak attempted to jab. If I hadn't moved to the side, it would have succeeded. I struck at its neck with the knife, but the feathers were too thick. I couldn't make contact with its skin.

"The other eye," Frall's scratchy voice called out.

I circled the bird, suspicious and ready, but it didn't move. I think it was dead. How could anything have survived losing so much body fluid? Yet, I wasn't positive, and its most recent attack had made me doubt its ability to die from such a wound. Thus, when I plunged the knife down in its other eye, although I normally would have been squeamish, this time, I didn't even pause. That bird/cat had wanted to kill us. I felt no compassion.

I stabbed and stepped back to avoid the flow of green, but little was ejected this time. Nor did the monstrous creature move or squeal out in pain. Bravely, I kicked at the limp body. Then, I reached out and touched the beak. It didn't move.

"It's dead," I told Frall, looking back at him. "I've done it."

I kicked at it again and looked back at Frall. "What now I asked? I can't pierce its neck."

Frall's eyes opened up, and he looked at me. I saw them widen before he let out a cry. "Clista!"

What I hadn't seen was the tongue sliding down the bird/cat's beak.

Instinctively, I jumped back, and the tongue, which had been springing for me, missed. With Frall's knife in hand, I pounced once more, cutting downward into the tongue's tannish flesh. Once more, green blood flowed, and I sliced clear through. The remaining tongue withdrew back into its mouth. The severed piece writhed.

"Dinner," Frall cried out, but the thought of it turned my stomach. I retreated, backing up, my eyes still watching the creature and that awful writhing tongue segment.

I don't know how the bird/cat had sent its tongue to attack that final time. The horrid creature looked even more lifeless by then. Yet, I didn't want to test the assumption. I simply watched and waited as it slowly dripped out the last drops of its greenish fluid.

It took several moments more, but then, at last, the bird/cat gave an enormous heave, a burp of stinky breath, and its wounded tongue dropped down, the beak opened, and, at last, it lay still.

"Proud," Frall said. "I'm . . . proud… of you."

I smiled, but Frall didn't see. His eyes had once again closed, his breath rasping each time he drew in another breath.

Darn, you, Thoric. Are you going to let Frall die? I cried out with a mental scream. Yet still, there was no answering murmur in my head, only silence.

I cleaned Frall's knife on a plump, moist leaf and then sat down beside him, watching the night's cover enfold us with its damp, cold blanket of dark.

The night was longer than any night had ever been. My eyes burned from staring into the gloom. They ached from sleepiness, too, but I held Frall's knife clutched fiercely, and I guarded the one I loved.

At some point, I must have drifted off. I don't remember ever closing my eyes, but I woke up on the same white clinical bed I'd been on before, the one in Thoric's hospital room. Frall lay on the bed across from me. I breathed a relieved sigh and fell back into a stress-free sleep.

Chapter Twelve

"Are you going to sleep around the clock?" Thoric whispered into my ear.

I jumped. The knife in my hand was instantly ready to fight.

"Ah, my warrior princess. How delightful. You did not achieve what I wanted, but I suppose you learned other things."

I can't tell you how angry Thoric's words made me. I leaped off the bed and stood in a crouched position, the knife in readiness for attack. Thoric just laughed.

"You are marvelous. What a beauty. What flaring eyes and heaving breasts you confront me with. I almost wish I were younger, my dear. What a fine queen you are becoming."

I jabbed at the air, wanting to see fear in his eyes, but he didn't back away from me. His eyes continued to laugh.

"Do you think you could frighten me, my dear? You? You are still an innocent baby. You wouldn't even make a decent challenge. Drop the knife, little one. Drop it before I am forced to demonstrate."

"Why do you want me to drop the knife if you're not afraid?" I said, taking a step forward.

"Stop playing with her," Nelona said from behind us. I was about to turn and include her in my challenge, but the knife suddenly flared with heat. I released it and yowled, shaking my hand from the sudden burn I'd been given.

"Nelona, was that absolutely necessary? Really. Come here, child. I will stop the pain."

I was teary-eyed from the ache of it. It felt like I'd been holding a red-hot iron against my palm. Yet, I stubbornly rejected Thoric's command. I was beginning to hate the *Old Ones*. So far, they had brought me nothing but anguish.

"Tirira," Thoric said. I had no idea what that meant, yet my feet brought me to him as if I were bound by a rope that he pulled hand over hand.

When I was close enough, he grabbed at me and studied my palm. "Ah, Nelona, the girl is in great pain. She is only a child, remember? We must go slowly with both of them. They haven't grown into their power, and they understand almost nothing."

My mouth opened for me to speak, but I couldn't issue the words. I felt lethargic, dreamlike. I almost floated instead of standing.

Thoric ran a small machine over my hand. It hummed like the larger one had done when examining my body. This one was faster. It bleeped after only a few seconds. Then, it sprayed something warm, coating my pain with relief. But that was only for a moment. The warmth turned icy cold. My hand felt frozen. I cried out, wanting to shake it, to rub it against my pants, to blow on it and warm it, but Thoric's grip didn't allow me to withdraw my hand.

"She fights even the healing. Do you not see the fear in her eyes? Perhaps we have tried to move too fast with them," Thoric said.

"They have our blood. They can endure it," Nelona told him, crossing her arms and striding forward to look at my hand. "Ah, it is nothing," she said. "This one needs to learn respect. She still doesn't know to obey the wisdom of her elders."

As if that were that, Nelona callously boosted herself up on a medical bed and sat beside me. "Both of them are ripe for enlightenment," she continued, lecturing Thoric. "It is only their stubbornness that prevents their instruction. We must persist. We must bring them to fruition. We have already waited too long."

"What have you done to my grandson this time, Thoric?" a new voice boomed from the open door. I recognized it as Goor's.

"He is sick," Thoric explained. "He picked up some kind of virus in the jungle. Luckily, she was immune, so the adventure was not all a waste."

Goor walked over to the machine next to Frall. He studied the data on it. "Why did you not pull them back sooner? You allowed this illness to weaken the boy's body. It invaded his lungs. He might have died. . ."

"You know what we wanted. It didn't happen."

"My granddaughter moved forward. That is what's most important," Nelona said, looking impatient with both men.

Meanwhile, the pain in my hand had lessened. I looked down and saw that it was covered with a foamy whiteness. I couldn't see the skin beneath. Were they restoring my hand as they had my arm?

Thoric observed the movement of my eyes. "Another moment," he said. "Then I'll wipe off the medicine, and your hand will be good as new, my queen."

"And I hope you have learned not to point weapons at us, Clista. Such instruction should not need repetition."

My mouth formed words, angry words, but my tongue was too stiff to utter it, my voice box unable to expel air.

"Will you send them back again when the boy is well?" Goor asked. "Stav mentioned that he wants to come see her. What should I tell him?"

"Neph, too? Is she coming here with him?"

"Of course, Nelona. When would Stav go anywhere without Neph?"

"She doesn't seem that eager to see what she has produced. I would have thought she'd come sooner."

"You know that she is in mourning."

"I'm in mourning, too. It was my daughter who was taken, but I have not given up on life. I am here, fighting for the child, attempting to meld her into a queen. Why is Nepth not doing the same?"

Goor laughed. "Does Neph really need to be here? It is obvious that you and Thoric have taken the lead. We others do not wish to interfere.

"Thoric, take the poor child out of *Tirira,*" Goor urged. "Her mouth is flapping open and shut. It's obvious she wants to speak."

"No, keep her in *Tirira*. She argues when she's fully awake. What's the point?"

"The point is that she is the queen, Nelona, not a slave miblira."

Nelona jumped down from the clinical bed, marched over to her husband, raised up her hand, and caressed his face. "Goor, you sweetie. Thank you for caring. She is a beautiful granddaughter, is she not? I am actually very pleased. And you? Are you not happy with your gallant grandson?"

It was the first time I had seen the two interact in a loving manner. I gaped, especially when Goor kissed Nelona right on the lips.

"Now, will you let her go, whichever one of you has put her under this?" Goor requested.

"In a minute," Thoric told him as he took a clean cloth from a drawer and wiped off the foam. "There, my queen. Good as new. Is that not better?"

Feeling Goor's eyes still on him, waiting for the release, Thoric sighed. "*Aririt*," in a voice that made it apparent that he wasn't eager to undo the spell.

"Thank you," I said, withdrawing my hand from Thoric's grip. "My hand feels much better."

"Did you learn your lesson, Clista?" Nelona asked impatiently.

"Yes," I said, staring down at the floor so she wouldn't see the fury in my eyes. "I'll never again wave knives at *Old Ones*."

"Are you positive this is my grandchild? It can't be. She has learned absolutely nothing," Nelona said angrily.

I sighed and squelched my anger. "Or other weapons," I added, a little more meekly.

"Give her time, Nelona, and be patient," Thoric told her, trying to soften her anger.

"Patience," Nelona snapped at Thoric. "I have been patient. That daughter of mine was totally unable even to see a dream sketch. She ignored every forewarning I sent her. She persisted in her research despite my admonitions. And look where it got her?

"Patience. I have no more patience. I have a granddaughter who is as blind as her mother…"

"Stop it. You can criticize me all you want, but leave my mother out of it," I told her, forgetting the caution I'd just learned.

"My mother wanted so much to 'feel' your artifacts. She was desperate to be the one the necklace called. With every fiber of her being, she desired that connection. But she couldn't. It wouldn't let her, not even to touch it. And nothing spoke to her. Nothing called."

"How ridiculous. Nothing called. Why do you think your parents discovered the find of their lifetime? Do you think that was luck, child?" Nelona asked, her eyes mocking my ignorance.

"You mean you guided her?"

"I made her fall down the incline. Of course. Do you think she would have sensed anything without that? She was as cold a receptor as your uncle and that man, Smog."

I took a step closer to my grandmother. "Then you did help her? You wanted her to find the artifacts?"

"I was her mother, Clista. Of course, I wanted her to be happy. I was proud of her, too, sometimes, although I would have preferred her to be different, more open, more imaginative. That woman who raised her was too distrustful of anything non-scientific. She stole away Clara's ability to believe, to be open enough to perceive something unknown to science. It was sad because that kind of training robbed your mother of much of her heritage.

"We thought your father would make the difference. He was open. He saw things she couldn't, but it was too late for her. Instead of helping her, she closed *him* off. Clara shut all his windows, and the rigid, scientific mind absorbed both of them."

Nelona sighed, wiped a tear, and continued. "But your parents gave us *you*. They shut you out often, but that was good because you grew up receptive. Let your skepticism go, Clista. Open up your receptors. Feel the power in the air and inside you."

Thoric cleared his throat, about to speak, but Goor beat him to it. "That is enough, Nelona. Now you've gone the other way and given her too much," he said. He took her arm and drew her closer to him. "Come," he said gently, "let us take the child down to the table and feed her. She looks hungry."

It was true that I'd had nothing to eat, and it had been several days. No wonder I felt sluggish and limp. Yet, I wasn't hungry. I just wanted to sleep again at Frall's side, my arms wrapped around him.

"I won't leave Frall," I told Goor. "I will eat when he does."

Nelona laughed bitterly. "Observe what I have been talking about. She still has not learned. Over and over, we have guided her and shown her, but still she doesn't see."

I wasn't sure what it was I was supposed to see. I glanced back at Frall. He was asleep. I sighed and looked over at Thoric. Then I blinked because the walls had blurred, and I was moving. Mists surrounded me. I believed another adventure awaited me. I groaned. Would I be alone this time? The thought frightened me. "Frall," I cried out.

But the motion stopped, and I blinked once more as the mists disappeared. I'd only been transported to the grand table down on the second floor.

"I told you I would not eat without Frall," I said angrily, and I turned to walk away.

"Must we take away your ability to walk?" Nelona asked.

I swallowed and looked back at Goor, who was shaking his head.

"Sit down," he ordered when he saw he had my attention. "Thoric is with your mate. He will watch over him. Surely you can be parted from Frall for the time it takes you to dine."

"Sit down, Clista," Nelona barked, impatience in every word and in her face. *Why is everything a battle with you?* she blared into my mind.

"If I answer, will you rob me of my voice or shift me outside?"

"Sit down and eat, and then I will allow you to talk. Later. Only do not argue with your elders, young lady. That is not seemingly."

Seemly? Where did Nelona get her vocabulary?

Grudgingly, I sat and immediately reached out for an apple. The table was spread with everything I'd dreamed of in the jungle. But now, it just seemed like overload, and I no longer felt any hunger.

Nelona ladled out a bowl of minestrone soup. Goor passed me a plate of hamburgers. My mouth felt dry. I reached for the cup that sat in front of me. It was filled with cocoa. Three plump marshmallows floated on top.

"You heard me, or you read my mind. Does that mean you knew the whole time how sick Frall was?"

Goor shook his head. "Not I, little one. I was far away most of that time."

"Thoric was the spotter. He watched you this time. It is he who selected the meal," Nelona said. "Pity. I would have chosen differently. Would you not prefer a lighter meal? Perhaps some cheese to go with the apple?"

I nodded. "Please. I know I said this is what I wanted before, but my stomach won't let me. I'm not really that hungry now."

"But you will eat. You need to build back your strength," Goor said.

I sighed and put down the apple, just as Nelona served my plate with several slices of white cheese."

"Why should I want to do that, Goor? Then you will send me back on another trip, more horrible than the last. I don't want to go again. I'm too tired to fight anymore. No more monsters. No more searching desperately for food and water. Please."

"Done," Goor said. "And that is final, Nelona. I will not allow the girl to be sent again in her worn down condition. Let her be a child for a few days. Let her play and read a book. There is no sense in driving her beyond her readiness, and it is obvious she is not mature enough to do what you really wanted."

Nelona stood up and glared at her husband. "How dare you interfere. This is my granddaughter. I will not abide your meddling in my training."

"You forget yourself, wife. She is my queen. I have the right to forbid it. I believe Bayn will back me up. He and Medior have decided to visit. Neph and Stav are on their way, too. Why should you have the only say in your granddaughter's upbringing?"

"Neph and Stav. They are my grandparents, right?" I asked, trying to get it right.

"Eat your apple," the irate Nelona ordered. "And the cheese, too." Then she sighed. "All right, Goor. You win," she said, sitting down in her chair. "When have you ever *not* meddled, my husband?" she asked, laughing dryly.

"Yes, child," Nelona said, responding to my question. "They are your grandparents, your father's biological mother and father. Perhaps you will bond with them. You have certainly not done so with me."

I looked down at my plate. I should say something, but the truth is that I didn't have any reason to like my grandmother.

Goor stepped in to smooth over the ugly moment. "Medior is your great grandfather on your mother's side. Bayn is your great grandfather on your father's side. He is Neph's father."

I couldn't keep all of that straight. I chewed my apple and nibbled at the cheese. "So, four more *Old Ones* are coming. They won't all be sending me places, will they?" I asked, coughing because the apple had gone down wrong.

"Haven't you learned something important on each of your trips? Haven't they been adventures?" Nelona demanded, as if I'd attacked something she believed in strongly.

"They've been horrible, each in a different way," I said truthfully. "The only one that was even halfway enjoyable was the forest. But then you separated me from two friends that I guess I'll never see again."

"Your friends in the forest? Silly girl," Goor said, grinning. "You're looking at them right now. "I was the jasmire, Merot. Dear Clista, I tried hard to protect you. I'm very sorry the wild jasmire attacked. I shouldn't have allowed you from my sight."

"You were Merot? That's not possible!" I said, bolting up. "You were a jasmire."

"Will she never learn? Not possible. Blah! Stop thinking like a human," Nelona said. "All things are possible for us. We long ago left the world of negativity. We travel where we wish, in whatever form pleases us. Haven't you learned that yet?"

I finished chewing my mouthful of apple, studying Nelona as I did so. Then I swallowed. "You were Yte?"

She nodded. "Rather slow analysis, wouldn't you say, Clista? Skip your mind over the logic part. Transverse the realms of all you believed was impossible, and you will see more clearly."

I saw that even as a jasmire, she had not much liked me, preferring the company of Frall. I wondered why. I didn't ask. I just sighed deeply.

"Eat your apple, Clista. Frall may not be as steadfastly patient with your immaturity if you grow too skinny for him."

"Do not attack her all the time, Nelona. She believes that you don't love her," her husband said.

Nelona's face turned red, and her eyes flashed at him. "Love. Blah. If Clista had the brains her heritage should have given her, she'd know that all my efforts are due to affection. It is the only reason I agreed to train her. I have no interest in her being our queen. The blood of family binds us. It calls us together. We are one."

Did I understand any of that? I was pretty sketchy about her reasoning, but I think she was trying to say that she did care about me. I sure wished she'd show it.

I stood, having finished the cheese and half the apple. I felt that if I touched another bite, I'd eject it. "May I go check on Frall now, please? I have eaten enough."

"You have eaten almost nothing," Goor said. "Finish the apple."

"No," Nelona told him. "She asked politely, dear. A woman's stomach cannot eat when it's upset. And she has broken her fast. Now, she will know her hunger when it comes. Let her go."

Goor sighed, but he bowed his head to his wife's will. Then he waved, and I felt the blurring, the mist, and the movement all at once. This time, before I could blink, I'd reached the hospital room. Slightly dazed, I waited a moment, allowing the mists to dissolve. Then, I took a step toward Frall.

"Did you eat something?" Thoric questioned.

"Obviously," Nelona told him from behind me. "Do you think I'd have permitted her to leave the table without having done so?"

"Of course. I did not mean to imply otherwise," Thoric wisely said.

"Has Frall awakened yet?" I asked. "His body needs food, too. Can you not…"

"Ah, listen to her. Now she has become a medic," Nelona laughed.

I sighed but continued watching Frall sleep. I couldn't see the rise and fall of his chest. He lay so still, it frightened me.

"Let us try. I'll allow you to feed him, Clista. If he isn't able, I'll hook up a drip directly into his blood, but it would be better . . ."

"Here," Nelona said. "I brought up some soup and sweetened juice. They will help him."

I walked toward Nelona, reached out my hands, and thanked her. I think she flashed half a smile at me, but it was gone so quickly I wasn't sure I'd really seen it.

The clinical bed was slowly raising Frall's head. As it elevated, Frall opened his eyes. He looked for me first. Seeing me standing beside him, he grinned.

"My brave . . .woman," he said.

I feared that his speaking would start up the coughing again, but it didn't.

"Did you bring . . us . . back?" he asked me, closing his fingers around the hand I'd placed on his chest.

I shook my head. "No. I tried, Frall. But it didn't work."

"Then — Thoric . . . ?"

Frall sighed and glanced at Nelona and at Thoric. "Thank you," he said. "I thought. I would not . . . make it . . .this time."

I'd pulled a chair over so that I could sit and spoon soup into Frall's mouth, but I was too low to reach him. I stood up.

"Sit down, Clista," Nelona ordered. "The chair will take you to the height you wish. Simply order it with your mind."

While I was attempting to adjust the height, Thoric moved closer. "I waited to give her a chance, Frall," Thoric explained to his grandson. "You understand that, don't you?"

I slipped in a spoonful of soup before Frall could respond. He smiled at me and swallowed it. "I'd rather . . have . . a drink," he said.

I handed him the cup, and he drank about half of it and then rested the cup on his chest. I took it back and set it on the table beside me.

Frall nodded. "Did you . . . give me . . . this. . . illness, . . . Thoric?"

The spoonful of soup stopped midway as I gasped and turned to stare at Thoric. I hadn't thought of that. Had Thoric really been that mean?

"No," he said, meeting both our eyes with an expression that held no guilt. "Nothing went according to my plan. I did not desire your illness, Frall. I thought the absence of food would be enough to force you to bridge the jump.

"And I didn't foresee the jasmire attacking you, Clista or that bird thing attacking Frall. You are not a food source for either of them. I don't understand what they thought would be gained by preying on you. I will be investigating that."

"I'm glad it wasn't one of you that I killed," I said thoughtfully. "That would've been awful."

Frall looked at me as if I were losing my mind. I had to stop and explain to him how Goor had just told me that he and Nelona had been the jasmires, Merot and Yte. It was obvious Frall hadn't realized that either. He shook his head. "Good thing . . . we didn't . . . know that… You might . . . have hesitated . . . dangerously."

"Stop talking, Frall. Please. You need to eat some more soup. I know you'll feel better after that," I urged.

Obediently, he opened his mouth for several more spoonfuls, but he, like me, had little appetite. He waved his hand that he was finished.

"A drink?" he asked.

Again, I handed him the juice. He finished the glass and closed his eyes.

Frall had dribbled juice down the side of his mouth. I wiped it off, then kissed his lips. He opened his eyes. "Don't," he said. "I'm probably contagious."

He turned to the side, and a second later, he was snoring softly, his chest rising and falling rhythmically.

"Didn't you say I was immune to whatever Frall has? I asked.

Thoric showed a moment's worry. "I suppose I had better check. I need to prepare a vaccine for it. I doubt such a virus can affect us, but it is always wisest to proceed cautiously."

"What about my uncle, Franca, and Smog? Where are they? Will you vaccinate them, too?"

"I wondered when you'd ask," Thoric said. "I placed them back into a long sleep. It's for the best. They interfere too much with what we're doing. They will be safe in their rooms," he said, stopping me before I could protest.

I hadn't been about to argue. What good would that have done? "But you'll vaccinate them even though they're asleep?" I asked.

Thoric smiled. "Yes, my queen. It will be done."

Thoric was studying his medical computer display unit. I went to stand behind him and watch.

"Sit down, child. Your body is still stressed from the adventure. I do not mind your observing me, however."

"Where is your wife, Thoric? Did you put her into deep sleep, too?"

Thoric stopped and looked at me. "No," he said. "She is in her room, not feeling well, I'm afraid."

"She doesn't have what Frall has, does she?" I asked worriedly.

Thoric laughed. "She carries our child, Clista. She is not sick like that."

His announcement amazed me. I'd thought he said the *Old Ones* couldn't have children in their own bodies. "Congratulations," I told him. "A girl or a boy?"

"Ah, here's the analysis. Now, we can determine the vaccine. A girl, Clista. One just like you, maybe," he said, smiling at me.

"You mean she'll be a queen, too?"

I watched as his fingers followed the series of letters on the printout, none of them legible to me.

"Why would you say that? There is only one queen. They come every three to five generations. You are she, Clista. There will be no other in your lifetime, and you will live very long."

I was always confused about this queen business. It seemed so strange that the *Old Ones* would call me their queen — me who'd never even been to Flagan Pren before.

"What does a queen do, Thoric? Is it like a job?"

"I'm sorry, Clista. I cannot talk about that with you," he said, looking uncomfortable.

"But shouldn't I know? Maybe I should be doing things that I'm not doing."

Thoric waved his finger at me. "No. You must find your own way. No one can help you with that."

"Except Frall?"

Thoric laughed. "Yes, except Frall. I hear your stomach growling again. It's time for you to fill it once more."

"I don't want to go down there again. Do I have to?"

Thoric tilted his head and stared at me. "Nelona bothers you? She *is* difficult at times. You will become used to her. She has many fine qualities, my dear."

Then he sighed heavily and said. "You will not find a better friend than your grandmother. What a shame there is so much friction between you."

"Robot, bring a tray of food for the queen. Deliver it to the medical lab," he ordered, looking upwards as if it was the air that took his orders.

"Some for me, too?" Frall asked.

"You're awake," I cried out and hurried to his side."

"And still contagious," he said, holding his hand up as if that could stop me.

I ignored it and hugged him. "Are you feeling better? Can you breathe okay now?"

"I hope that vaccine is ready. She never listens," Frall told Thoric, but I noticed that Frall was already pulling me closer, not pushing me away.

"It will be ready before I send her off to bed."

"I'm not leaving Frall alone," I stated, ready to argue.

"Don't make it harder on me, Clista," Frall whined, releasing me. "I'll sleep in my own room and you in yours — unless you've decided to marry me."

"You're well enough?"

"For marriage or to go to my room by myself?"

The robot entered. I turned to look, partly out of curiosity, and partly so I wouldn't have to answer Frall's question. In so doing, I saw that Thoric's eyes were examining me as well. He had raised up one eyebrow in question, waiting for an answer.

Luckily, food temporarily took Frall's interest away from the question. I removed a plate from the robot and delivered it to Frall.

"Thanks, darling," Frall said as he pushed himself up. He didn't need me to feed him anymore. I could tell that he felt better because he gobbled up his hamburger before I'd even taken a bite of mine.

I chewed mine slowly, attempting to do as Nelona insisted, to let go of preconceived ideas.

I loved Frall. That was not my barrier. It looked like I wasn't going back to Rwawan. There'd be no more Academies, no more chatting with friends in the quad, or being just a teenager. There was only the here and now. Things were different in Flagan Pren. Maybe I should cave in and marry Frall. The *Old Ones* wanted me to. Everyone did, except my uncle and Smog.

I was still three months from a majority, or was I? Could I count the time I'd been in a deep sleep? That lasted a year. If I added that into the equation, I was already of legal age. Should deep sleep count?

"Clista, wake up," Frall kidded me.

I smiled at him and finished my hamburger. Then I drank my juice and stood up, ready to return our dishes to the robot. I collected Frall's and started to walk off.

"Clista, it's all right, darling. I do understand," Frall said, grabbing my arm. "Take your time. I didn't mean to pressure you. When it's right, the words will fly off your lips, and you'll *want* to marry me, not feel you *have to*."

I concentrated on looking at his hand on my arm. I couldn't bear to meet his eyes. "I do love you," I said.

"Yeah, I know." He wiped away the tear sliding down my cheek. "We'll sleep in our own bedrooms, Clista. That's how it has to be. You know that, right?"

I nodded. "But it will be hard."

Frall laughed. "It will be easier than those nights when I lay there watching you sleep, wishing…"

He released my hand, swung up out of the bed, and stood up. Then he clutched at the metal of the bed, holding on as if he'd collapse without its steadying frame.

"Let me help you," Thoric called and rushed over to take Frall's arm. I followed behind, dumping the dishes onto the robot's outstretched arms.

That night Frall and I did not spend back to back. As he'd demanded, we went to our separate beds. The mattress, or the pithor, which is what the *Old Ones* called the softness inside the bed platform, was far more comfortable than the damp-leafed ground. The blankets on top of me were soft, warm, and dry, but my sleep that night was fitful. I very much missed Frall's gentle snores and the warmth of his breath on the back of my neck.

Chapter Thirteen

In the morning, two more *Old Ones*, ones not even expected, arrived — Pekl and Brest. They were Frall's great-grandfathers. So he was well represented with heritage. On my side, I still had only Nelona.

We dined with them, and for some reason, they studied me more than Frall. I guess I examined them, too, so maybe we were equal. Luckily, though, they didn't test me as Goor had done. For that, I was grateful.

Pekl was taller even than Thoric and didn't look old at all. His hair was slightly reddish, almost auburn. Did men have auburn hair, or was that a woman's term? I couldn't remember.

Pekl was handsome. Frall looked a lot like him, much more than he resembled Thoric. I could see Frall's smile when I looked at Pekl.

Brest, on the other hand, looked grandfatherly. He had white hair — a full head of it, and his eyebrows were thick, white, and bushy. His smile was completely different from Frall's, but I liked him anyway, especially when I overheard him tell Pekl that I was not only beautiful, but smart, since I was treading cautiously in the new world I'd found myself.

I wasn't sure what he was referring to, but anyone who calls you beautiful AND smart deserves a special place in your heart.

Oh, did I mention that Frall was well enough to come down for breakfast? It was really exciting to see him looking healthy again. I kept staring at him and smiling with happiness. I'm sure that everyone thought I had a super bad crush on him all of a sudden.

Pekl kidded me about it, telling Frall how lucky he was to have a "pretty, young girl" doting on him. (That wasn't my word, but Pekl's.)

Frall didn't act very pleased about it. He was sitting next to Brest, who I think had said something Frall didn't like because Frall ended up with that expression he gets when he's mad about something. At one point, Frall even glared at me.

I mouthed a "What's wrong?" but he shook his head. I knew that Frall could speak into my mind. He did that sometimes. I didn't understand why he wouldn't answer me.

Before Frall had finished the food on his plate, he and Brest stood up and walked away from the table — without a word to the rest of us. I thought about joining them, but the truth is, I hadn't been invited. Frall hadn't even taken a moment to look me in the eyes or let me know in some other way that he wanted me to accompany him.

I sighed and tried to listen to Pekl telling me about Stafle, the planet he'd just been visiting. Normally, that would have been interesting, but I was worried about Frall and about how he'd walked off after giving me a rather displeased look.

Nelona started telling everybody about the jasmire attack. Breena came downstairs right in the middle of the story. She didn't look good. I hoped she was feeling better than the day before. Someone had once told me that women got sick when they were pregnant. Had my mother? Had Nelona?

Breena didn't eat anything, but she drank some weak tea. Then Thoric moved his chair closer, and he whispered into her ear. She gave him a smile of such tenderness I almost cried. It reminded me of how my father used to look at my mother and how sometimes she used to lay her hand on his shoulder and just leave it there.

Dad would always reach up and cover her hand with his. It had looked awkward to me. I'd always wondered why they did it. Why didn't they just hold hands like other people? But it was just something they used to do. I guess it didn't seem awkward to them at all, or else my father wouldn't have repeated it so often.

My parents used to have names for each other. My father's real name was Felix, but my mother had called him "Cat." I once asked her about it, but she just laughed. Dad called Mom "Meow" because he said she was the "cat's meow." I figured all parents were sometimes just weird like that, but now, watching Breena and Thoric, I wondered if it maybe had something to do with being married. Perhaps married people just got quirkier.

Would Frall and I have our own gestures and make up names for each other? I was thinking about that when I heard Frall's voice in my mind. *I already have a name for you, darling. "Mermaid," remember?*

I turned to look, but I couldn't see Frall. Wherever he'd gone, he wasn't in the same room as we were. Still, I smiled. Frall wasn't angry with me after all. I brightened up considerably and tuned in to hear Pekl talk about what people were wearing on the planet, Stafle.

Later, when we rose up from the table, Nelona took charge of me. She escorted me up to the garden room via the conventional method of elevator transport. The garden room was the place where Frall and I had found the fountain — the one we'd played in that day.

I wished I could take my shoes off and dip my feet in the water, but when I asked Nelona if I could, she lectured me about protocol. Apparently, it was not appropriate for a queen to go around barefoot.

It was a good thing Nelona hadn't paid any attention to me when I was growing up. I'd spent most of my playtime barefoot.

"Sit down," Nelona told me, and there was suddenly a bench where, a moment before, there hadn't been one.

"Now, you and I are going to have a chat," she said with her strong, forceful voice. "This time do not force me to send you away. I want you to behave yourself."

If I'd dared, my mouth would have dropped open. Me, misbehave? As I recalled, I hadn't done anything before when she'd sent me away.

"Later today, well, actually, at any moment, your relatives will be dropping in. You are to be cordial to them. Understand?"

I nodded. I was always polite when people didn't do strange things to me, like sending me off to desert windstorm environments.

"Medior and Bayn — those are your great grandfathers. They will demand to know about what you will do as queen. Ignore their questions. Do not be rude, of course, or you might find yourself in a situation and location much worse than any place I sent you. Bayn once got angry with someone and sent him to his death."

I gasped. "You mean he killed someone?"

"No. Well, yes, but then he brought the fellow back to life. I gather it was most uncomfortable, however. Dying is rarely pleasant. You certainly do not want to antagonize Bayn."

"What are you telling my great granddaughter, Nelona? Nonsense? She has no reason to fear me. I would never send my own flesh and blood to death, not even for a brief experience. Besides, she is much too pretty to endanger.

"Come, child. Give your old grandpa a kiss."

"Great Grandpa, you mean, Bayn." Nelona corrected him.

Unlike Nelona, I figured this was not the time to quibble over titles. I stood up and walked into his arms. His hug was a bit fiercer than I would have liked, but he kissed my cheek and seemed pleased with me. When I was free, I looked up at him and smiled a bit nervously.

"Sit down, girl," Bayn said with a deep, gruff voice. He created another bench for us, one much bigger and more comfortable than Nelona's.

I sat down a bit gingerly where he'd ordered me to sit, casting a worried look at Nelona. I certainly didn't want to make *her* angry with me.

"You have her scared of you already," Bayn snorted and then began to chuckle like a rusty steam kettle.

"Oh, are we having a party?" asked a couple, dropping down through the ceiling as if they'd been walking on the roof and merely took a wrong step.

Nelona jumped up. "Nepth, Stav, how wonderful to see you! How well you look."

Bayn grunted as if he thought otherwise.

Not knowing what was expected of me, I stood up.

"This is the child? Why she's so young. I thought the way you talked that she was a good twenty or thirty years old. How old are you, child?"

"I think I'm seventeen, but I might be eighteen. Does deep sleep count for determining one's age?"

All four *Old Ones* laughed. "She's adorable," said Neph.

"She certainly has the look of her father. I'd know her anywhere. Come here, child," ordered Stav.

I walked toward them, suddenly shy. I'd been trying to get used to being around *Old Ones* for several days, but to encounter four of them at the same time, three of them unpredictably new, was quite overwhelming. What if they all decided to send me someplace at the same time? Would pieces of me go in different directions? Would one leg end up at the beach and another in the desert? Or would one of them override the other? Would I survive a battle as they argued over my destination?

Bayn started laughing. The others almost immediately joined him. Then, my great grandfather began to cough. I turned to see if he needed my help, even though I had no idea what to do about it.

"She thinks we're going to tear her into pieces," Bayn hooted. "Can you believe that?"

Had they all been reading my mind? I backed up from Bayn and turned to look at the new couple. Their eyes told me that they were privy to my vision, also.

"Must you read my mind?" I flared. Then I darted a quick look at Nelona. Had I angered her? It was then I remembered what she'd said about Bayn. The fountain was looking more and more inviting. If I dived in and held my head under the rapidly moving water, would they still see me?

"Come here," Stav commanded again, and this time, it was the same voice that Thoric used when he was giving me no choice in the matter. I walked toward him, not stopping to consider the wisdom of doing so.

When I was close enough, Stav reached out and pulled me the rest of the way, his hands gripping my shoulders firmly. "Ah, Clista, do you know how happy we are to meet you face to face finally? We have longed to do so for many years."

"We have watched you grow," Neph added. "I was there the day you fell from the tree. Do you remember that?"

The tree? Suddenly, I did recall that moment. There'd been a lady there that day who'd comforted me. She'd put a bandage on my knee. I'd been crying because I'd fallen. I hadn't meant to cry. Mother never allowed me to do so. But I'd been frightened, and my knee had hurt so much. I was worried that I'd done something awful to it. I couldn't even stand at first, and then the lady came with the bandage, and I was suddenly okay again. Why had I forgotten that?

"Look at her expression," a strange voice called out. "That's her mother's habit."

My nose must have crinkled up like it did sometimes when I grew puzzled. Frall was always kidding me about it. I was never aware that I was doing it, but I'd certainly seen my mother do it many times. I swiveled my head to see where the voice came from, but whoever had spoken was not only silent but invisible.

"Medior, come out of hiding," Nelona demanded.

"Daughter, must you spoil my fun?" a head asked as it popped into the room. The head rotated and came closer until it was hanging in the air, staring at me.

"So you are the big hoopla," the head said, and a hand came out of the air to join the head. The hand was pointing at my nose. "That's my family's nose, all right."

Do you know what I wished for at that moment? For a math test. Even if I didn't understand the theories that the proctor was testing me on . . . it didn't matter. The electronic pencil in hand, the tablet where I bubbled in the answers, which were immediately marked correct or incorrect — all that would have been a relief. The hard, straight chair, the beveled desk, the clock ticking away on the time limit. Oh, sweetness. Comfort. Normality.

"Sit down, child. You look pale," Neph said.

"Of course, she's pale. Look at what that idiot is doing, parading like a *Snizvor* in midair. What is the girl supposed to think?" Bayn fussed.

Frall, I whispered in my mind. *Are you busy?*

Do you need me? he asked, sliding into my head comfortably, soothingly.

I don't know how he did it, but a second later, Frall was sitting next to me on the bench that Nelona had made, his arm around me. He pulled me close.

"In case you didn't know," he told them, "my name is Frall."

The head turned and winked at me. "She called you, did she? Well, I'm delighted with that. Wonderful news if you ask me."

Stav said, "I'm glad to finally meet you," and held out a hand to shake Frall's. "You have done a good job of imprinting her."

Imprinting? I shot a glance up at Frall's face. "Imprinting?" I asked.

"I'll explain later," he whispered, and he gave my forehead a quick kiss as if that would relieve me from worrying over Stav's word choice.

"Careful," Nelona warned Stav. "You will find that Clista is quick of mind and skittish as a *Varmilli Trestom*."

"Let me see if I have this correct," Frall smoothly injected an interruption. "You two must be Clista's grandparents, uh, Neph and Stav, right?"

They nodded. Bayn sat up straighter. "Good thinking, boy. But who am I? I am certainly not one of the grandmothers. You can see that," he said, laughing even though his words were not in the least bit amusing.

"You are either Medior or Bayn, which makes you Clista's great grandfather. Let's see. I think you look more like Neph than Nelona. I would guess you're on Clista's father's side."

Bayn let out a huge guffaw. "Right again. I certainly am not on Nelona's side," he chuckled. "I do not have the right head for it," he said, chuckling over his pun.

"Is he making fun of my head, daughter?" Medior asked, joining in with a laugh.

Nelona sighed and rolled her eyes. "Honestly, the two of you are enough to send the children straight back to Rwawan. Settle down, both of you."

"Oh, let them have their fun, Nelona. At least Jow is not here. That would be really bad."

"Where is he?" Medior asked. "That boy of mine is never where he should be. How dare he not be here when we need him!"

Frall was whispering in my ear, o "Are you okay, now? Nothing like having all your relatives drop in at once."

Nelona told Medior and the others where Jow was. I'd never heard of the guy. But whoever he was, I was glad he wasn't coming. There were already too many *Old Ones* here. I couldn't keep them all straight.

Thank you so much for coming, Frall. I was about ready to dive into the fountain and stay there — underwater for the next couple of days, I told him, laughing.

It was my pleasure. You know that, mermaid. I'd always rather be with you than Brest.

I sighed and darted him a kiss. Could anyone be luckier? Frall was so incredibly sweet.

"Hey, you two, pay attention. We were just asking you something," said Medior, the one who had a decapitated head and an occasional arm.

Neph laughed. "Oh, be still, Medior. I think the two of them are adorable together. It makes me smile to watch them."

Stav frowned. He was like a vid star, one with a smile that made young girls sigh. His hair was rather longish and night black. His eyes beamed with blue lights that sparkled in an eye-catching manner. His nose was noble, his forehead regal. He was, I sighed, very, very handsome.

Lost in the analysis of his face, I almost didn't hear his words. "They would be adorable, Neph, if they would marry as they were meant to. I do not understand their procrastination. It is her duty to bring forth an heir."

Had I said handsome? I changed my opinion immediately. An heir? I didn't even know if I was ready for marriage yet, and this guy was demanding that I have a baby? No way. He wasn't that handsome. He was too cold to be handsome.

Easy, Clista. Remember what they can do to us, Frall whispered into my mind.

Stars, he was right. I didn't want to be sent to another monster-filled island or a sandy desert where the wind attacked with sandpaper force. I did some meditative breaths and focused on relaxing.

"Give her time, dear. She is still a child. We have waited two hundred years. We can wait a few more."

"Not many," Bayn laughed. "Some of us do not have that many years to wait."

My eyes moved to Neph. I decided I liked her. "Thank you, Grandmother," I said.

"Hush, you will call me Neph. You will call all of us by our names. We do not use relationship names as you did on Rwawan. We are aware of relationships without titling them."

Neph was smiling at me. Her fair, young face was the perfect complement for Stav's darkness. Her golden hair fell down to her shoulders and had gentle waves of sunlight in them. She was as beautiful as Stav.

If they were really my father's parents, what had happened to my father? He'd been okay looking, but not as glamorous as they were. It was kind of amazing to be related to them.

"If I were the boy, I'd know how to get down to business. You wouldn't even have to send me to an island to do it," Medior said.

Frall got very quiet. I could tell that he was embarrassed. He'd stiffened, and his face had turned several hues of red. Why couldn't the *Old Ones* leave our relationship out of their plotting? Neph had tried to help, but I was suddenly angrier than before.

"It's none of your business what Frall and I do," I said.

Frall groaned.

The *Old Ones* stared, their faces displaying gasps of surprise.

"I warned you," Nelona said, the first of them to recover.

"Let her be. I like her spirit," Medior defended me.

"She *is* full of sass. Must take after me," Bayn laughed.

"It cannot be allowed," Stav stated.

"Dear, remember when . . ." The rest of that was lost to us because Nepth was whispering in his ear.

Frall shook his head at me. *I thought you didn't like deserted islands with monsters, darling. I think you've just managed to get us sent to another one.*

I suppose that might have happened, but Thoric popped in, joining the *Old One's* convention. "No, I put my foot down," he said firmly, looking each one of them in the eye and ending up at Nelona, whom he stared at until she finally dropped her eyes.

"Frall is absolutely not ready to journey again. He was very, very sick, and I am not convinced that his lungs have recovered." Thoric told them. "Give the boy a couple of days to strengthen his body. In fact, I'd recommend a week."

Nelona started to speak, but Thoric cut her off before she could say a word. "No, Nelona, I do not advise separating them at this stage. They are bonding. Do not send the girl off without him. Do not."

"They look pretty well bonded to me," laughed Bayn. "You could not get them any closer together if you glued them," he snorted loudly, staring pointedly at Frall's arm around me.

"That does not mean they are bonded," Medior's head said. "Seems to me that when I was their age…"

"Now, really, you two. Stay relative to the manner at hand, please," Nelona snapped. "So, if they should not be separated, Thoric, and you say Frall is not well enough to be sent out for a few days, what in the *scmisph* are we supposed to do with them? It is obvious the girl needs a lot of coaching. She is a dreadfully slow learner."

Of course, I didn't like that. My own grandmother — and she said that in front of everyone. My face burned. I wished once again that I could dive into the fountain, except this time, I wished it led to another place, preferably a planet where there were no *Old Ones*.

Clista, ignore her. There's nothing the least bit slow about you. Just say nothing, Frall sent.

"Why, Nelona, I cannot believe your words. It was my understanding that Clista scored extremely well on all the tests given to her during her deep sleep time. Is that not true, Thoric?" Neph asked.

Stav gave Thoric a puzzled look. "Yes, I read all the data on her. What is Nelona talking about? The girl scored superiorly in all categories."

"You have not been here observing her these days on Flagan Pren. Her stubbornness and Rwawan's primitive viewpoints have interfered with almost everything we've tried to teach her. She is stuck in a holding pattern and is afraid to come out of it."

I'd had enough of this. Despite Frall's hold on me, I stood up, opened my mouth, and started to tell them what I thought of their lack of manners.

I had just opened my mouth and pushed out the words, "This is the ru…" when Frall started coughing and bent over with a horrifyingly ominous groan.

"What's the matter?" I said, forgetting my whole tirade. "Frall, are you okay? Are you sick again?"

Bayn burst out with a laugh. The others either stared at Frall or at Bayn. We were all confused — Frall doubled over with pain, while Bayn chuckled away like that was a good thing.

"Stop it," I yelled at him. "Can't you see that Frall is ill? How can you laugh at a time like this?"

I was so angry I could have thrown something at the *Old One*. I didn't care what Bayn might or might not do to me. He was horrible and mean!

Bayn glanced at my expression, took in my words (and probably my thoughts), and instead of being angry, as I expected him to be, he burst out with another horrible snort and guffaw. "She does not know. She does not know."

Of course, I didn't know. Why didn't he tell me if there was something to tell me? My eyes circled the others at the same time as I was putting my arms around Frall and bending over him. "What can I do to help, Frall?" I whispered. "Tell me, please. You know I'll do anything."

"Anything?" he asked with a terrible twinkle in his eye.

"You're not sick, are you?" I bolted up in outrage. "Are you telling me that was just an act? Were you faking it?"

He whispered something that I didn't hear. I wasn't sure I could trust him after that stunt, but I bent over closer to hear what he had to say.

"Clista," he whispered as he reached out to clasp my face in his hands. "I'm sorry, darling. I know I shouldn't have done that, but you were going to pop off with something acid-strong, weren't you? Bayn has a reputation. Brest was telling me about that while I was downstairs. I didn't have time to warn you before. But if he takes issue with something, his retribution for the offense is severe. Please think about that. Do you really want to find out what an angry *Old One* can dish out?

"Remember our class? Remember what we learned about letting go of rage? Think about it for a moment, darling, *please*."

He leaned forward and kissed me on the lips, but I was still angry. I would have pulled away, but he released me before I could jerk free. Then he grabbed my hand.

"Clista, no matter what you do, I'm still on your team, and I'm not oppositional this time like I was when the proctor at school forced me to be. I will always be on your side from now on. I don't have any choice about that because, you see, no matter what happens, I love you."

Then he dropped my hand and stood up. I stood up, too, and stared up at him. Everyone was watching us and waiting to see what would happen next. I wanted to turn and look at them, but I couldn't yet. I couldn't take my eyes off Frall's.

He used to be one of the cutest guys at school. He had a smile that devastated, a brain that impressed, a way with words that kept all the girls listening to his every uttering. But he wasn't that same guy anymore.

I blinked as if to bring him back. Where was the boy I'd once sighed over? Where was the student who'd angered me, then befriended me, and later earned my heart? He was gone, and in his place was a man who comforted, who counseled, who poured out his heart, and who had become my rock of constancy.

"Wake up, Clista," Frall said, teasing me gently.

I blinked. The man I'd seen a moment before was still there. "You've changed," I said softly. "When did you change?"

He stared at me for a moment like I was crazy. "Are you okay, Clista? What do you mean?"

Temporarily, no one was bothering us. In the back of my mind, that worried me. The *Old Ones* were probably having an internal conversation and excluding us from it. That thought irritated me, but it didn't distract me enough from trying to understand my new eyes, or at least what it was that I was seeing with them.

Almost shyly, I placed my hand on Frall's shoulder. "It's just strange. I kept seeing you at school. . . I mean, the way you were at school. . . but that's all changed, hasn't it? We're not those people anymore."

Frall studied me. Was he seeing the same kind of changes? Had I aged as he had done?

"What do you mean, Clista? Changed — in what way?"

He sat back down on the bench, his legs spread apart. He reached out with his hands and pulled me in closer. Then his arms folded me into a kind of embrace that was as much about the intensity of his focus as it was the tenderness of his cuddling.

"Tell me what you're seeing, Clista. Please," he said.

I blushed. How could I? Wouldn't it insult Frall to tell him that I'd thought of him as a teenager, an adolescent — but never as an adult, a grown-up man?

I didn't have to put it into words. Frall smiled down at me, and then he kissed my forehead. "I know, Clista. It's what you wanted to see. It's what you needed me to be. That's why I never pushed you."

"But you're different now. Suddenly."

He didn't say anything. The pupils of his eyes dilated. It was like he was looking into me with a stronger, better lens.

"Who am I now, Clista?"

I was a coward. Once again, I wanted to back away and run to hide in the shallow water of the fountain.

If you swim into the depths, I would still follow you, my mermaid. I would follow you no matter where you went.

"And you'd wait forever?" I said, pulling the words from something he'd said before.

"No. I won't wait forever, Clista. I won't need to wait forever. You're almost there now, aren't you?"

It was one of those moments that should never have been interrupted. It was the climax, the bridge we needed to cross to reach the other side. Would I have taken it? Would I have been brave enough to meet Frall in the middle?

But Nelona didn't give us the opportunity. "All right. I give in," she yelled. "All of you are united against me. Let her have her way, but I am warning you . . ."

"Oh, stop with your dire predictions. Look at the child, Nelona. How much longer can it take?" Neph asked.

Frall and I lost our concentration. The *Old Ones* had come back into the equation, interfering, tossing in words that irritated and separated. It was like having ice cubes dropped down our backs. Knowing that the *Old Ones* were centered on us completely disconnected us from wherever this new awareness might have taken us.

"Well, I have not changed my mind," Stav said. "She should be punished for her impertinence."

Neph put her arms around her husband's neck. "Do you not think that Frall will be able to change her attitude better than we could, dear husband? You certainly changed mine."

Stav glared at me once more, but then his gaze softened as he stared into his wife's beautiful eyes.

He sighed. "What do you think, Frall? Can you tame her?" he asked.

That was the kind of talk that had incited me the first time. I took a deep breath and exhaled it slowly. Frall watched and nodded, smiling at me.

"I think all the excitement has tired out Clista," Frall said. "She was exhausted from her last challenge and has not had time to recuperate. I think it would be better if she be allowed to rest now."

If Frall had said that at any other time, I would have been furious, but I understood that he was buying us time. I nodded slightly, just enough for him to see. He swung his arm around my shoulder and started walking me forward.

"He is right. She was fatigued by it," Thoric agreed. "She killed one of the beasts, you know. It attacked, and she was incredibly courageous. Clista probably saved Frall's life."

"Is that right?" Medior said, getting so excited about it that his body showed momentarily from the waist up.

Frall and I took advantage of their distraction, scooted by them, and then continued on, strolling contentedly into the apple orchard.

"They can talk about that for a while without us," Frall said.

"You know what's funny," I said. "A long time ago, before she flew to Crexkor, I had a best friend named Sandra. She had five brothers and a sister and used to talk about how they drove her nuts. I was always jealous of her. I wanted brothers and sisters, too. But you know what, Frall, if all these relatives are what it's like having a big family, then, I'm glad I don't have siblings. What about you? Did you sometimes want to have siblings?"

"I wasn't an only child, Clista. I used to have a little sister. She got sick and died when she was only five."

I stopped. "Died? I'm so sorry, Frall," I said, taking his hand. "What happened?"

"Remember the Curlian Plague?"

I shook my head. I'd never heard of it.

"Settlers brought that back from one of the new planets. It lived in spores. We didn't have a cure then. The Curlian Plague took out a lot of kids, at least in Thepra, where I lived.

"My parents and I moved into the city afterward. Dad and Mom wanted to get away from their memories. Trena — that was her name; she left memories everywhere. Every place we went, every tree, every flower in the yard, and every piece of furniture. We started fresh. We took almost nothing with us."

I threw my arms around Frall and held him. I couldn't imagine losing a little sister. How terrible! How sad. "How old were you at the time? I mean, when she died?" I asked after a moment of silence.

"Ten, so it was eight years ago if you don't count the deep sleep time."

"I'm so very sorry. Tell me what she was like, unless that bothers you too much."

Frall remained silent for a moment and then began to speak hesitantly. "She was like you. What you must have been like at age five, anyway. She was brave, too brave sometimes. She climbed every tree in the yard. And she was constantly asking questions. That's the part you that you and Trena shared. I can see you at age five doing that. You still ask a lot of questions, you know," he said, smiling into my eyes.

"What else?"

"Well, she used to drive me crazy 'cause she was always getting into my stuff. One time, I found her trying to read one of my textbooks. Of course, she could barely read, but she was studying the pictures and trying to learn. She was that smart. And she was a quick learner. If time had given her another year, I bet she would have read that book and understood it, too. I still miss her. Clista. I miss her a lot."

I sighed and wiped a tear. I was such a jerk not to have ever asked. I'd just assumed he'd been an only child. "Thank you for telling me about her, Frall. It makes me feel special that you trusted me with your memories."

"All my memories are yours, darling. All you have to do is reach inside me. I'm always open to you."

I knew what he meant, but I didn't feel right about doing that. It felt strange tapping even the outside of his thoughts. I shook my head. "I'm not ready for that."

"I know. Patience. Now, where were we before the *Old Ones* interfered? You were telling me how I'd changed, grown older. Tell me more about that, Clista. I don't want to talk about my sister anymore. I want to talk about us."

We had stopped in a patch of grass. Frall pulled me down and slid me next to him. "No, that isn't right," he said. "I want to see your eyes, Clista. Sit opposite me."

He swung my legs over his. I wasn't comfortable with that, but my uncle wasn't awake, and we were just talking.

His eyes glinted, smirking over my shyness at such things. "Okay, now. Talk. Tell me why you think I've changed. How am I different, my lovely mermaid?"

He picked up my hands and held them. I watched him weave my fingers into his. Even his hands had matured, growing longer, bigger. Didn't he see that? Wasn't he aware how much his body had physically changed since we'd come to Flagan Pren?

"Look at your hands," I told him. "Back at school, they were just a little bigger than mine. Now they're huge."

Frall laughed. "Yeah, I noticed the growth spurt. Good thing Thoric supplied us with new clothes. Nothing would have fit if I had to wear the stuff in my old closet."

"I thought you'd already reached your height on Rwawan. You were almost as tall as my father. Now you're bigger than he was."

"It happens like that, Clista. My father was big. I'm the same age as when he hit his growth spurt. You might grow a little more, too, but I doubt it. I think you'll always be on the small size."

"Thanks," I told him sarcastically.

He chuckled. "I like your size, Clista. I think you're perfect. You know that," he said, leaning forward to kiss me.

"Where could they have gone? I saw them head this way," came a familiar voice. Both Frall and I groaned. Nelona had found us.

She stumbled into our miniature glen and stopped to stare at us. "Did you not hear me calling you?" Her voice, as usual when she spoke to either of us, was angry. She turned to face the way she'd come. "Thoric, I found them. They were clumped together, kissing."

With faces red from embarrassment, Frall and I stood up and waited for the lectures.

"You children were rude to wander off," she started in at once. "Your family is here, Clista. You owe them respect. You must return immediately."

"But I told them Clista was tired and needed rest," Frall explained. "No one said anything when I took her away from them. She needs…"

"You were not resting when I walked into your little enclosure," Nelona said crossly. "Come, Clista. No more excuses."

Thoric caught up with Nelona just in time to hear her words. "Easy, there," he cautioned her. "Never push a queen too hard. She may discover resources she does not yet know exist."

Thoric's eyes were staring into mine. I knew he was sending me a message, but I had no idea what he was talking about. Believe me, I would have resisted Nelona's bossiness a long time ago if I'd had any special queenly resources.

"Bah," Nelona said, almost spitting with disgust. "They're locked inside her. She spent too long with her parents being indoctrinated in that scientific method of narrowmindedness."

"Leave my parents out of your criticisms," I snapped.

Be cautious, Clista, please, Frall warned, inserting the thought silently into my mind.

"Are you scolding them again?" Medior asked, popping into the air. This time, all we could see of him was his eyes and half of his nose. His nose wasn't flesh-colored as before, but gray as if he'd turned into a rain cloud.

"She is always scolding us," I mumbled, ignoring Frall's advice.

Frall threw his arm around me and turned me to face him. Crushing my head against his chest as if he were hugging me, he said, "Clista and I were just discussing how we'd always wanted more family. It is so nice that you were able to travel here to meet us."

"Nice?" Medior said, laughing, his eyes darkening into pellets of black. His head developed a bit more — a white wreath of cloud mist. "We are not here to be nice. We have come to push Clista into her birthright."

"Something she is apparently not interested in," Nelona told him.

"Not interested? Does she have any idea how fortunate she is? We have planned for her many centuries now. How dare she not rise to the occasion."

As if to give credence to his sudden temper, Medior sent a flash of light, let out a huge crack of thunder, and rained down on the ground.

"Behave yourself, Medior. Don't get us wet," Nelona snapped, jerking her long, flowing turquoise cape well away from the area that was being watered.

I tried to speak, but Frall was being irritatingly restrictive. I raised my knee warningly.

"Darn it, Clista," he snapped, sounding almost exactly like Nelona, except three octaves lower.

I didn't like his style, but I knew he was attempting to protect me. I couldn't fault him for that. My mouth did fly loose with things it shouldn't be saying. I laid my hand on the thick of his arm and said, "Sorry."

It was enough. He smiled down at me and offered his hand. I took it and turned to face the *Old Ones*.

"Cleverly done. But it's good to see that she hasn't allowed the big guy to dominate her," Medior said.

"Of course, she has. She has permitted far too much of his authority. Why we…"

"I take issue with that, Nelona," Goor said, popping into the scene. He landed next to Nelona and almost immediately swung his arm around her and pulled her against him. "Since when have you not listened to my council and been guided by it?"

"That is different," she said, suddenly flustered. "You are. . ."

Take note of this. It is very important, Frall spoke into my mind. *What is it that harnesses the argumentative Nelona?*

I wasn't sure what Frall meant by that. I turned my head to look up at him. His eyes were twinkling. Heeding Frall's words, I watched the couple more intently. They were speaking mind to mind; that was obvious. So, how could I observe anything?

I looked up to find Thoric's eyes on me. He nodded and motioned to Goor. I was still mystified until Medior chuckled and said. "Yep, I chose wisely when I paired those two. Nelona cannot push him around. That guy is loaded with power."

Power? Okay. I had pieces, but as if it were a puzzle without a picture or a frame, I didn't know where anything went. Too many pieces and no clues.

I disconnected myself from Frall and walked over to Thoric. Nelona and Goor were still deep in conversation, although we couldn't hear a word. I spoke quietly so only Thoric could hear. "So, Goor is stronger than Nelona in power? How does that help me? I don't understand."

"You came to the wrong person for understanding, Clista. I told you once before that Frall has your answers."

"But he doesn't know either, and you do. If I have power or Frall has it, how do we harness it? Why wasn't I able to transport us back here? I tried so hard. Why did I fail? What's the key?"

Thoric took my hand and led me back to Frall. "Take her somewhere else and talk with her. Explain. I'll see if I can keep the others from bothering you, two. No promises how long, though."

So once more, Frall and I moved off to escape the *Old Ones*. When we found another quiet spot and were leaning against an apple tree, munching its yellow-gold fruit, I opened up the conversation. I told Frall about my failure to bring us back to Thoric's and how I'd tried everything, including the necklace.

"Thoric's right, Clista," Frall said, brushing a strand of hair from my eyes. "I do know the problem, or should I say, the solution."

I shot up from my position and kneeled at Frall's side so I could look into his eyes. "What is the solution, Frall? Tell me. I tried everything."

He wasn't looking into my eyes. He avoided them. He threw his unfinished apple off to the side. "Calm down, Clista. Give it time. You heard Neph speak of how we're still young…"

"You're stalling, Frall. Spit it out. What is the great mystery here that everyone keeps avoiding telling me? No more secrets, please."

He sighed. "Clista, power requires us to get closer, a lot closer, if you know what I mean."

"Closer?" Then I blushed. I knew what he was saying, and I suddenly understood what everyone had been talking about with their veiled looks and their allusions.

"I don't believe that. It shouldn't make any difference. That's ridiculous…" I sputtered.

"It isn't a reason to rush into marriage if you don't want to, Clista. But it is the truth, at least what Thoric and the others believe."

I sat down again, my back against the tree. I didn't touch Frall. I didn't want to. I wanted to think about what I'd just learned and to put the pieces all together.

"That's why they've done what they've done to throw us together so many times. They thought we'd..."

"I've never liked being pushed, Clista. And no one is going to throw us together when we're not ready for it.

"You know, kids used to have intercourse before they were ready . . . but it wrecked their lives. They couldn't handle it. They found out that it destroyed love more times than it strengthened it. I don't want to do that to you. I'll wait, my darling."

I knew the history, too. We'd learned it on Rwawan. No dating while we were at the Academy, and then only sparingly. We'd all attended classes that explained how long ago kids were short-changed. Their dreams were crushed by early pregnancies or emotional depths that sputtered and died, often sending teens down into a spiral of depression.

I'd never intended to form a relationship with Frall, not until I'd finished schooling and was company hired and solidly into my future. But then, somehow, we'd been pushed into it. Was it our heritage, our destiny, or the *Old Ones'* meddling? And did it really matter why or how? It was done now. We were a couple. That was indisputable.

"Yes, it matters," Frall said, picking up my hand and taking it to his lips. "It matters because you need to know yourself first. You need to figure out what you want, what is right for you, and for us.

"We are the new. We can write our own path, Clista. We can choose to follow the customs of Flagan Pren, the *Old Ones' beliefs*, or even Rwawan's, if we wish. Or we can write our own.

"We can do that together, and all your relatives and mine can't really have a say in it. They can lecture and send us off. They can put your uncle in a deep sleep, but they can't change who we are, Clista, or what we think is best for us."

I sighed. "But they have power. You heard Thoric. We have no power without doing as they wish."

"No, Clista. We have far more power than that. We have the power of saying 'no' and holding on to our beliefs. We have the power of our love, and there's nothing in any world as strong as love."

I threw my arms around Frall and kissed him for being so smart and because I did love him more than anything in the whole world — in the whole universe. And that kiss was like a pact between us. It was our acceptance that we could wait and not let the pressures of the *Old Ones* rule us.

We stood up, hand in hand, and together, we returned to our relatives. We were stronger and firmer then and resolved to build the future in the way we wanted it, in the manner that was best for *us*.

At least, we thought we could . . . But our adventures don't end, and the *Old Ones* don't give up.

Will we gain our power? Will I truly become the queen that everyone envisions me to be?

Read more of Frall and Clista's story in:

Visions of the Future

Book Three of the Alien Dream Series

Made in the USA
Monee, IL
04 November 2026

67381204R00134